Lecture Notes in Computer Scienc

Commenced Publication in 1973
Founding and Former Series Editors:
Gerhard Goos, Juris Hartmanis, and Jan van Leeuwen

Mark Claypool   Steve Uhlig (Eds.)

Passive and Active Network Measurement

9th International Conference, PAM 2008
Cleveland, OH, USA, April 29-30, 2008
Proceedings

 Springer

Volume Editors

Mark Claypool
Worcester Polytechnic Institute
100 Institute Road, Worcester, MA, 01609-2280, USA
E-mail: claypool@cs.wpi.edu

Steve Uhlig
Delft University of Technology
Mekelweg 4, 2628 CD Delft, The Netherlands
E-mail: s.p.w.g.uhlig@ewi.tudelft.nl

Library of Congress Control Number: 2008924790

CR Subject Classification (1998): C.2, C.4, H.4, K.6.5

LNCS Sublibrary: SL 5 – Computer Communication Networks and
Telecommunications

ISSN   0302-9743
ISBN-10   3-540-79231-7 Springer Berlin Heidelberg New York
ISBN-13   978-3-540-79231-4 Springer Berlin Heidelberg New York

Springer is a part of Springer Science+Business Media

springer.com

© Springer-Verlag Berlin Heidelberg 2008
Printed in Germany

Typesetting: Camera-ready by author, data conversion by Scientific Publishing Services, Chennai, India
Printed on acid-free paper   SPIN: 12257515   06/3180   5 4 3 2 1 0

Preface

The 2008 edition of the Passive and Active Measurement Conference was the ninth of a series of successful events. Since 2000, the Passive and Active Measurement (PAM) conference has provided a forum for presenting and discussing innovative and early work in the area of Internet measurement. PAM has a tradition of being a workshop-like conference with lively discussion and active participation from all attendees. This event focuses on research and practical applications of network measurement and analysis techniques. This year's conference was held in Cleveland, Ohio.

PAM2008's call for papers attracted 71 submissions. Each paper was carefully reviewed by at least three members of the Technical Program Committee. The reviewing process led to the acceptance of 23 papers. The papers were arranged into eight sessions covering the following areas: addressing and topology, applications, classification and sampling, measurement systems and frameworks, wireless 802.11, tools, characterization and trends, and malware and anomalies.

We are very grateful to Endace, Intel and Cisco Systems whose sponsoring allowed us to keep low registration costs and also to offer several travel grants to PhD students. We are also grateful to Case Western Reserve University for sponsoring PAM as a host.

April 2008

Michael Rabinovich
Mark Claypool

Organization

Organization Committee

General Chair   Michael Rabinovich (Case Western Reserve University, USA)

Program Chair   Mark Claypool (Worcester Polytechnic Institute, USA)

Publication Chair   Steve Uhlig (Delft University of Technology, The Netherlands)

Program Committee

Mark Allman	ICSI
Grenville Armitage	Swinburne University of Technology
Surendar Chandra	Notre Dame
Jae Chung	Airvana
Wu-Chang Feng	Portland State University
Wu-Chi Feng	Portland State University
Paal Halvorsen	University of Oslo
Manish Jain	Telchemy
Simon Leinen	Switch
Kang Li	University of Georgia
Ibrahim Matta	Boston University
Ketan Mayer-Patel	University of North Carolina
Anees Shaikh	IBM Research
Colleen Shannon	CAIDA
Augustin Soule	Thomson Research
Peter Steenkiste	Carnegie Mellon University
Ooi Wei Tsang	National University of Singapore
Steve Uhlig	Delft University of Technology
Carey Williamson	University of Calgary
Craig Wills	Worcester Polytechnic Institute
Huahui Wu	Google
Michael Zink	University of Massachusetts

Steering Committee

Mark Allman	ICSI
Nevil Brownlee	University of Auckland
Mark Claypool	Worcester Polytechnic Institute
Ian Graham	Endace

Konstantina Papagiannaki   Intel Research Pittsburgh
Michael Rabinovich   Case Western Reserve University
Matthew Roughan   University of Adelaide
Steve Uhlig   Delft University of Technology

Sponsoring Institutions

Endace
Cisco Systems
Intel Corp.
Case Western Reserve

Table of Contents

Malware and Anomalies

The Flattening Internet Topology:
Natural Evolution, Unsightly Barnacles or
Contrived Collapse?

Phillipa Gill[1], Martin Arlitt[1,2], Zongpeng Li[1], and Anirban Mahanti[3]

[1] University of Calgary, Calgary, AB, Canada
[2] HP Labs, Palo Alto, CA, USA
[3] IIT Delhi, Delhi, India

Abstract. In this paper we collect and analyze traceroute measurements[1] to show that large content providers (e.g., Google, Microsoft, Yahoo!) are deploying their own wide-area networks, bringing their networks closer to users, and bypassing Tier-1 ISPs on many paths. This trend, should it continue and be adopted by more content providers, could flatten the Internet topology, and may result in numerous other consequences to users, Internet Service Providers (ISPs), content providers, and network researchers.

1 Introduction

Since its creation in 1969, the Internet has undergone several significant changes. From its beginnings as a research network, the Internet evolved into a commercial network by the mid-1990's [5]. The emergence of "killer applications" such as the World-Wide Web and Peer-to-Peer file sharing vastly expanded the Internet user base [11]. For a variety of reasons, including the commercialization and increased popularity of the Internet, it has become extremely difficult to make ubiquitous changes to the Internet infrastructure. This has led to the emergence of *architectural barnacles* [15], or ad hoc work-arounds for a variety of architectural problems. Architectural purists argue that barnacles may provide short-term relief to such problems, but over the long-term only exacerbate the underlying issues [15].

In this paper we examine a new trend that is emerging at the infrastructure-level of the Internet: large content providers are assembling their own wide-area networks. This trend, should it become common practice, could result in significant changes to the structure of the Internet as it exists today, and have numerous ramifications for users, ISPs, content providers, and network researchers.

We find that companies such as Google, Yahoo!, and Microsoft, are deploying large WANs. Google is leading the way, with a WAN infrastructure that covers much of the U.S., and extends to Europe, Asia, and South America. Yahoo! and Microsoft also have WANs covering the U.S., but do not (yet) extend to

---

[1] Our data is available at the Internet Traffic Archive - http://ita.ee.lbl.gov/

M. Claypool and S. Uhlig (Eds.): PAM 2008, LNCS 4979, pp. 1–10, 2008.

other regions of the world. These efforts may force other Internet companies to follow suit, in order to remain competitive. For example, MySpace appears to be partnering with Limelight Networks, a Content Delivery Network, to build out a WAN for MySpace.

Our paper makes several contributions. First, we alert the network research community to this emerging trend, as it may affect the assumptions used in other studies. Second, we provide initial measurements on the number and size of the networks already in place for some large content providers. Third, we describe the potential implications of this trend, and discuss whether this is a natural evolution of the Internet architecture, an unsightly barnacle which will ultimately create additional problems, or a contrived attempt to disrupt the balance of power among the providers of the Internet architecture.

2   Background

2.1   Internet Architecture

The Internet architecture has evolved throughout its history. Initially, a single backbone network connected a small number of research networks, to enable researchers to remotely access computing resources at other institutions [5]. In the late 1980's, commercial ISPs began to form, and by 1995 the backbone network was completely transitioned to commercial operation [5]. This transformation resulted in the current three-tiered organization of the Internet infrastructure: backbone networks (Tier-1 ISPs), regional networks (Tier-2 ISPs), and access networks (Tier-3 ISPs) [5,11]. Consumers and *content providers* access the Internet via Tier-3 ISPs. A Tier-2 ISP connects a number of Tier-3 providers to the Internet. The Tier-2 ISP peers with other Tier-2 ISPs to deliver their customer's traffic to the intended destinations. Tier-2 ISPs may also connect to some Tier-1 ISPs, to more directly reach a larger fraction of the Internet. There are only a few Tier-1 ISPs. Tier-1 ISPs *transit* traffic for their customers (Tier-2 ISPs), for a fee. Tier-1 ISPs peer with all other Tier-1 ISPs (and do not pay transit fees) to form the Internet backbone [11].

2.2   Motivations for Change

There are a number of reasons why content providers may be motivated to build their own wide-area networks, rather than utilize ISPs to deliver content to users. Three broad categories are *business reasons*, *technical challenges*, and *opportunity*. We discuss each in turn.

When the "dot-com bubble" burst (around 2000), many Internet companies, including Tier-1 ISPs such as WorldCom, Genuity, and Global Crossing went bankrupt [13]. This *economic collapse* [13] motivated surviving (and new) Internet companies to increase their focus on "business essentials", such as *risk mitigation* and *cost control*. One risk mitigation strategy content providers may employ is to reduce their dependencies on partners. This could avoid disruptions in a content provider's core business, if, for example, a partner declared bankruptcy. Similarly, topics such as "network neutrality" may create uncertainty for content providers, and hence motivate them to build their own WAN infrastructures, to mitigate any possible or perceived risk. To control costs, a company may look for ways to reduce or eliminate existing costs. One strategy for content providers is to utilize settlement-free *peering* arrangements with ISPs, rather than traditional (pay-for-use) *transit* relationships [14]. For large content providers and small ISPs, peering can be a mutually beneficial arrangement.

Content providers may also be motivated to build their own WANs for technical reasons. For example, a content provider may wish to deploy a new "killer" application, such as video-on-demand. Although many scalable video on-demand delivery techniques exist, none have been widely deployed, owing to the lack of IP multicast on the Internet. This limitation is due to the "Internet Impasse" [15]; this predicament makes it nearly impossible to adopt ubiquitous architectural changes to the Internet that might improve security, enable quality-of-service or IP multicast [16]. A private WAN could avoid this impasse, and give content providers more control over their end-to-end application performance.

Some companies, such as Google, Yahoo!, and Microsoft, aim to provide "Software as a Service" (SaaS), which will deliver functionality via the Web that was previously only available through software installed on the user's computer. In response to the shift to SaaS, several companies are making multi-billion dollar investments in infrastructure such as large data centers [6,12] and WANs. The motivations for these investments likely span both the business and technical categories described above.

Lastly, content providers may be motivated to build their own WANs because of opportunities that arise. For example, due to the bursting of the "dot-com bubble", a content provider may be able to inexpensively obtain WAN infrastructure (e.g., installed fiber optic network links) from bankrupt ISPs.

3 Methodology

3.1 Data Collection

Our measurement of the popular content provider networks utilizes the `traceroute` tool. `traceroute` is a tool that is commonly used to identify network topology.

To determine the extent of content provider networks, we decided on the following data collection methodology. First, identify a set of N popular content providers. For each of these content providers, select an end-point (i.e., a server). Next, select a set of M geographically-distributed nodes to issue `traceroute` queries, to gather topology information. Lastly, issue $N \times M$ `traceroute` queries. It is important to note that in this study we are only interested in identifying the end points of content provider networks; we are not trying to measure the end user experience, as this would require a different methodology (since end user requests are typically redirected to nearby servers).

For this study, we collected a single snapshot of the networks of the 20 top content providers, as ranked by Alexa [1], by querying from 50 different `traceroute`

Table 1. Top 20 Content Providers, as Identified by Alexa.com

1	www.yahoo.com	6	www.myspace.com	11	www.hi5.com	16	www.friendster.com
2	www.msn.com	7	www.orkut.com	12	www.qq.com	17	www.yahoo.co.jp
3	www.google.com	8	www.baidu.com	13	www.rapidshare.com	18	www.microsoft.com
4	www.youtube.com	9	www.wikipedia.org	14	www.blogger.com	19	www.sina.com.cn
5	www.live.com	10	www.facebook.com	15	www.megaupload.com	20	www.fotolog.net

servers. The 20 top content providers we used are listed in Table 1. We believe this snapshot is sufficient for an initial view of these networks.

We resolve the hostnames of the popular sites only once, and only at a single location (the University of Calgary). We believe this approach will prevent our queries from being redirected to local instances of servers. Since our goal is to understand the size of content provider networks, and not to measure the end-user performance, we argue that our approach is reasonable.

Although we only selected 50 nodes to issue queries from, we selected the locations of these nodes such that they are (potentially) biased in two ways: towards the country in which the content provider is based; and towards areas with higher concentrations of Internet users. We argue this is reasonable as we expect content providers will expand their networks to areas with the largest numbers of (potential) users first. At the time of our study (September 2007), 15 out of 20 of the top global sites listed by Alexa were U.S. based. As a result, we selected 20 `traceroute` servers in the U.S. These servers were located in 20 different states, including the 10 most populous states. 18 of the U.S. based `traceroute` servers are at commercial sites, and the other two are at universities. The remaining 30 `traceroute` servers were selected from countries around the world. Although we intended to use the 30 countries with the most Internet users, some of these countries do not have public `traceroute` servers. Instead, we issued queries from two locations in Canada (a workstation at our university, and a public traceroute server at another) and from 28 additional locations from around the world, in countries which had working public `traceroute` servers listed on `traceroute.org`. Overall, the 30 countries (including the U.S.) we selected were among the top 40 countries in terms of most Internet users, according to Internet World Stats [10]. The 30 countries we used account for an estimated 82.7% of all Internet users.

To keep the load on the 20 selected servers low, we issued only a single `traceroute` query from each server to each destination, and only one query at a time. Furthermore, we throttled the rate at which the queries were issued (this is in addition to throttling done by some of the `traceroute` servers). Our data collection occurred between September 27 and October 1, 2007. In future work, we plan to collect data periodically, to understand rate of expansion of content provider networks.

3.2   Data Analysis

In order to analyze the `traceroute` data, several challenges had to be overcome. First, automating the parsing of the data was problematic. Among the 50 different `traceroute` servers there were 10 different output formats. Thus, a parser was needed that could handle all of these. Second, the `traceroute` output only contained a portion of the data of interest. This meant it was necessary to find additional sources of data (e.g., IP address to organization mappings, organization to Autonomous System (AS) number mappings, etc.) Lastly, there were no obvious metrics for quantifying the size of the WAN of each content provider; this meant a lot of manual inspection of the data was needed in order to determine what the (automated) analysis should evaluate.

We overcame the first two challenges by developing a program to parse the outputs of the various traceroute servers. This program extracts the sequence of IP addresses for each of the traceroute queries. Once the sequence of IPs for a traceroute query is extracted, additional data about each of the IPs is gathered. First, the identity of the organization that registered the IP address is queried from the regional Internet registries. Second, the AS number for the IP address is resolved using an AS number lookup tool [21]. Gathering this extra information increased the potential analyses that we could perform on the data. Specifically, we were able to identify which of the hops in the traceroute path belonged to Tier-1 ISPs using a list of the nine Tier-1 ISPs and their AS numbers [23].

We selected four metrics to facilitate the comparison of the content provider networks, and to examine whether the Internet topology is flattening. We use the *average number of hops on Tier-1 networks* as a measure of how involved such ISPs are in the path. A related metric is the *number of paths that involve no Tier-1 ISPs*. Our third metric, which we call *degree*, provides a conservative estimate of the number of different ISPs a content provider is connected to. This examines the AS number for the router that immediately precedes the first router belonging to a content provider, on each distinct path. Lastly, we consider the *number of geographic locations* in which a content provider's routers appear to be located. We acknowledge that all of these metrics have their shortcomings. For example, it may not be meaningful to compare hop counts when examining differences in the paths. Hu and Steenkiste [9] describe similar issues for identifying metrics for comparing the similarity of end-to-end Internet routes. However, we believe our metrics nevertheless provide some interesting insights. For example, with the traditional Internet model we might expect popular content providers to peer exclusively with a number of Tier-1 ISPs, to ensure global coverage with a minimal number of exchanges on each end-to-end path. If, however, the Internet is flattening, we might expect to see more extensive peering with lower tier ISPs.

4   Results

In our analysis we observe that some companies own multiple top 20 sites. Specifically, we observe that Orkut and Blogger are both owned by Google, and traffic for these sites is carried on Google's network. We observe a similar trend for the sites owned by Microsoft, namely MSN and Live. Paths for all four of these subsidiary sites is carried on the same network as their parent companies, and thus the results are very similar. As a result, we only consider one site for each company when the traffic is carried on the same network. Therefore, for our results we omit Orkut, Blogger, MSN and Live, and only show the results for Google and

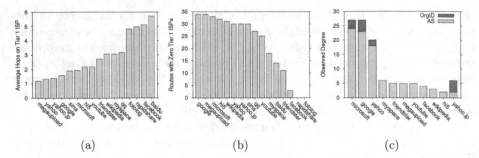

(a)   (b)   (c)

Fig. 1. Comparison of Network Paths to Top Sites: (a) Average Tier 1 Hops per Path; (b) Number of Paths with No Tier 1 Hops; (c) Connectedness of Each Site

Microsoft, the parent companies. Although Google has recently acquired YouTube, traffic for YouTube has not yet (completely) migrated to Google's network. Thus for our study, we consider YouTube separately from Google. Also, Yahoo! Japan has a unique AS number, so we consider it separately from Yahoo!.

Figure 1 shows the results for three of our metrics. Figure 1(a) shows the average number of hops on a Tier-1 network, for each of the sites. The most notable observation is that our `traceroute` probes traversed significantly more Tier-1 hops on average for some sites than for others. The more established "big three" content providers (Microsoft, Yahoo!, Google) were among those with the lowest averages. Figure 1(b) shows the number of (`traceroute`) paths to each site that contained no Tier-1 hops. For some content providers, including the "big three", 60% (30 paths out of 50) or more contained no Tier-1 hops. Figure 1(c) examines the degree of *connectedness* for each of the content providers that have their own AS number. This graph reveals a clear distinction between the "big three" and the other content providers. Our `traceroute` results show that Microsoft connect to at least 24 different ASes, Google to at least 23, and Yahoo! to at least 18. The next highest is MySpace, at only six. Some paths included IP addresses that we were unable to map to an AS number. For these IP addresses only, we used the organization identifier (OrgID) as retrieved from the corresponding Internet registry. This method enabled us to identify an additional three connection points for Microsoft (27 in total), four for Google (27), and two for Yahoo! (20). The only other content provider affected by this issue was Yahoo! Japan.

Figure 2 shows the geographic distribution of entry points into the WANs of selected content providers. Figure 2(a) shows the location of entry points across the U.S. The figure reveals that Microsoft, Google, and Yahoo! all have networks that span the country. The entry points appear to be located (as one would expect) in large centers where carrier hotels or Internet Exchanges exist. Google has the most extensive (live) WAN of any of the content providers we examined. Entry points into Google's WAN are shown in Figure 2(b). Our probes entered the Google network in 10 different North American cities, as well as four European, two Asian, and one South American location.

Other than the "big three", we did not detect any other content providers with large network infrastructures. For example, we only saw Facebook connect

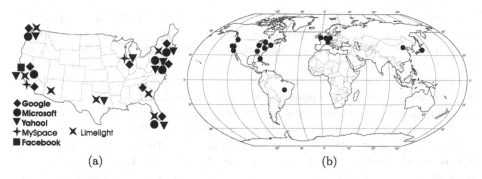

(a)   (b)

Fig. 2. (a) Location of network end-points in the United States for selected content providers. (b) Our measurement of Google's current WAN.

to ISPs in the San Francisco Bay area. We did, however, observe several things that suggest others are rolling out WAN infrastructures, in different ways. First, MySpace is peered with ISPs in two separate locations (Los Angeles and Chicago), and appears to partner with Limelight Networks, a Content Delivery Network, to reach other locations. Of 14 probes we sent from European `traceroute` servers to MySpace, eight entered the Limelight network in Europe (in Amsterdam, Frankfurt, or London), which entered Limelight's U.S. network in New York. Six other probes from different locations traversed the Limelight network in the U.S., before reaching MySpace. Second, YouTube (recently acquired by Google) appears to peer with numerous ISPs around the U.S. (We also noticed signs that YouTube's traffic is migrating to Google's infrastructure.)

5   Discussion

In this section we consider the potential ramifications of the identified trends. We discuss these from the perspectives of content providers, users, and ISPs.

If content providers build extensive network infrastructures, they could reap a number of benefits. In particular, they could gain greater control over network-related issues that affect their business. They could deploy applications that have been stymied by the "Internet Impasse". For example, there are reports that Google will deploy (or has deployed) computation and storage resources at the edge of their network [4]. This could enable Google to provide true video on-demand, and participate in the cable television market. Similarly, they would reduce their reliance on external providers, who might wish to compete against them. There are also many disadvantages. Perhaps the most significant is the cost of deploying, operating and maintaining the infrastructure. Although a few large content providers may have the funds to attempt this, it will be difficult for a large number to follow. In addition, as large content providers move their traffic off the (shared) Internet, small content providers may be faced with larger bills, if ISPs need to recover lost revenues. These issues may lead other content providers to re-examine cost/control tradeoffs; e.g., using VPNs rather than deploying physical networks.

Users could benefit from this trend in several ways. First, these "private" networks may provide better quality of service than the existing Internet, since the content providers could optimize their networks for the applications and services they provide. Second, users may get access to new applications and services much sooner than if they need to wait for a large number of ISPs to agree on a common set of supporting technologies to deploy. Over the long term, however, users could suffer if small content providers are unable to survive, as creativity may be stifled and the variety of content may decrease as a result.

Tier-1 ISPs may notice the greatest changes from this trend. In particular, if this trend becomes widely adopted, Tier-1 ISPs may need to adapt (e.g., vertically integrate, offer content services of their own, or implement new network functionalities that content providers desire, such as IP multicast), or face bankruptcy as revenue dries up. However, since large content providers are unlikely to carry transit traffic, the need for Tier-1 ISPs may not disappear. In fact, a possible (and counter-intuitive) side-effect of large content providers moving their traffic to private networks is lower costs for Tier-1 ISPs, as they may not need to increase the capacity of their networks as often (assuming large content providers are responsible for a significant fraction of the volume of Internet traffic). At the "bottom" of the hierarchy, competing with the "last-mile" ISPs (Tier-3) is unlikely to be attractive to content providers, as the last-mile is expensive to install, and the Return-On-Investment (ROI) relatively low. However, nothing should be assumed; Google recently qualified to bid on wireless spectrum in the United States, which could be interpreted as an initial step in providing last-mile wireless Internet service.

Our data suggests that the Internet topology is becoming flatter, as large content providers are relying less on Tier-1 ISPs, and peering with larger numbers of lower tier ISPs. Content providers are clearly exploring an alternative; only time will determine if this "mutation" becomes the new "norm", or an "abomination" which will eventually die off. However, this remains a hypothesis, as our results provide only two certain answers: (1) several large content providers are indeed deploying their own networks, and (2) it will be necessary to perform a more rigorous and longitudinal study to determine whether this trend is a short term *barnacle* (e.g., as inexpensive dark fiber disappears, will the trend end?), a slow, but certain evolution of the Internet (e.g., if greater peering between content providers and small ISPs occurs, the Internet topology could flatten), or a contrived collapse (e.g., content providers cunningly defending their territory against ISPs who wish to move into seemingly more profitable content services).

6   Related Work

Our interest in this topic was piqued by an article on a telecom news site [17]. This article stated that Google is building a massive WAN, and speculated that other large "Internet players" are likely doing the same. Thus, we wanted to determine if companies like Google have operational WANs, and if so, how large they are.

We are not aware of any work that has examined this specific trend. However, there are numerous prior works on tools, methodologies, or Internet topology measurements that we leveraged, or could leverage in future work, to answer the questions of interest to us. We describe some of the most relevant works below.

In this study we utilized `traceroute`; however, it has a number of known weaknesses [5]. Tools such as `tcptraceroute` [22] or Paris Traceroute [2] could be used in conjunction with PlanetLab to address these known limitations of `traceroute`. Sherwood and Spring propose additional methods for addressing these weaknesses [18].

The closest work to our own is Rocketfuel, which created router-level ISP topology maps [19]. A key difference is their paper focused on mapping the network topologies for specific ISPs, while we are interested in the network topologies for specific content providers. Spring et al. also proposed scriptroute, a system to conduct network measurements from remote vantage points [20]. Given the similarity in objectives, we will likely revisit Rocketfuel and scriptroute in the future. Additionally, scalability and efficiency of collection will be important for larger and repeated data collection efforts. Donnet et al. [8] and Dimitropoulos et al. [7] have investigated these issues for topology discovery.

A number of papers have discussed the need to evolve the Internet architecture, and proposed ways in which change could be enabled within the current (static) architecture [3,15,16]. In this paper, we examine a change that is occurring in the Internet architecture. Depending on how this change is viewed (e.g., is it a fundamental shift, or just an unsightly barnacle), it may be necessary to revisit the predictions of what the future Internet will look like.

7   Conclusions

In this paper, we utilized an active measurement (`traceroute`-based) approach to demonstrate that large content providers are deploying their own WANs. We show that established companies such as Google, Microsoft, and Yahoo! already have sizable WAN infrastructures, and find that some smaller (but very popular) content providers appear to be following their lead. While there are many possible motivations for this trend, we believe it is more important to consider the potential ramifications. Specifically, it could alter the way in which the Internet operates, either (eventually) eliminating the need for Tier-1 ISPs, or forcing such ISPs to evolve their businesses. Network researchers also need to understand whether this is a long or short term trend, as it will affect the importance of research topics.

Significant work remains to be done on this topic. Increasing the breadth of the study, conducting a longitudinal study, and considering alternative metrics are some of the dimensions of our future work.

Acknowledgements. The authors greatly appreciate the providers of the public traceroute servers as well as the feedback of Bala Krishnamurthy, Dejan Milojicic, Jeff Mogul, Carey Williamson and the anonymous reviewers.

References

1. Alexa's Top 500 Sites, http://www.alexa.com/site/ds/top_500
2. Augustin, B., Cuvellier, X., Orgogozo, B., Viger, F., Latapy, M., Teixeira, R.: Avoiding Traceroute Anomalies with Paris Traceroute. In: Internet Measurement Conference, Rio de Janeiro, Brazil (2006)
3. Clark, D., Wroclawski, J., Sollins, K., Braden, R.: Tussle in Cyberspace: Defining Tomorrow's Internet. In: ACM SIGCOMM, Pittsburgh, PA (2002)
4. Cringely, R.: Google-Mart: Sam Walton Taught Google More About How to Dominate the Internet than Microsoft Ever Did (November 17, 2005), http://www.pbs.org/cringely/pulpit/2005/pulpit_20051117000873.html
5. Crovella, M., Krishnamurthy, B.: Internet Measurement: infrastructure, traffic & applications. Wiley & Sons, Ltd., West Sussex, England (2006)
6. Data Center Knowledge Web site, http://www.datacenterknowledge.com/
7. Dimitropoulos, X., Krioukov, D., Riley, G.: Revisiting Internet AS-level Topology Discovery. In: Passive and Active Measurement, Boston, MA (2005)
8. Donnet, B., Friedman, T., Crovella, M.: Improved Algorithms for Network Topology Discovery. In: Passive and Active Measurement, Boston, MA (2005)
9. Hu, N., Steenkiste, P.: Quantifying Internet End-to-End Route Similarity. Passive and Active Measurement, Adelaide, Australia (2006)
10. Internet World Statistics (statistics from June 30, 2007), http://www.internetworldstats.com/
11. Kurose, J., Ross, K.: Computer Networking: A Top Down Approach. Addison Wesley, Boston, MA (2008)
12. Mehta, S.: Behold the server farm. Fortune magazine (July 26, 2006), http://money.cnn.com/2006/07/26/magazines/fortune/futureoftech_serverfarm_fortune/
13. Norton, W.: The Evolution of the U.S. Internet Peering Ecosystem. Equinox White Paper (2003)
14. Norton, W.: A Business Case for ISP Peering. Equinox White Paper (2001)
15. Peterson, L., Shenker, S., Turner, J.: Overcoming the Internet Impasse through Virtualization. IEEE Computer (April 2005)
16. Ratnasamy, S., Shenker, S., McCanne, S.: Towards an Evolvable Internet Architecture. In: ACM SIGCOMM, Philadelphia, PA (August 2005)
17. Raynovich, R.: Google's Own Private Internet, http://www.lightreading.com/document.asp?doc_id=80968
18. Sherwood, R., Spring, N.: Touring the Internet in a TCP Sidecar. In: Internet Measurement Conference, Rio de Janeiro, Brazil (2006)
19. Spring, N., Mahajan, R., Wetherall, D.: Measuring ISP Topologies with Rocketfuel. In: ACM SIGCOMM, Pittsburgh, PA (August 2002)
20. Spring, N., Wetherall, D., Anderson, T.: Scriptroute: A Public Internet Measurement Facility. In: USENIX Symposium on Internet Technologies and Systems (USITS), Seattle, WA (March 2003)
21. Team Cymru IP to ASN Lookup page, http://www.cymru.com/BGP/asnlookup.html
22. Toren, M.: tcptraceroute, http://michael.toren.net/code/tcptraceroute/
23. Wikipedia article, Tier 1 network, http://en.wikipedia.org/wiki/Tier_1_network

Assessing the Geographic Resolution of Exhaustive Tabulation for Geolocating Internet Hosts

S.S. Siwpersad[1], Bamba Gueye[2], and Steve Uhlig[1]

[1] Delft University of Technology, The Netherlands
mail@swiep.com, S.P.W.G.Uhlig@ewi.tudelft.nl
[2] Université de Liège, Belgium
cabgueye@ulg.ac.be

Abstract. Geolocation of Internet hosts relies mainly on exhaustive tabulation techniques. Those techniques consist in building a database, that keeps the mapping between IP blocks and a geographic location. Relying on a single location for a whole IP block requires using a coarse enough geographic resolution. As this geographic resolution is not made explicit in databases, we try in this paper to better understand it by comparing the location estimates of databases with a well-established active measurements-based geolocation technique.

We show that the geographic resolution of geolocation databases is far coarser than the resolution provided by active measurements for individual IP addresses. Given the lack of information in databases about the expected location error within each IP block, one cannot have much confidence in the accuracy of their location estimates. Geolocation databases should either provide information about the expected accuracy of the location estimates within each block, or reveal information about how their location estimates have been built, unless databases have to be trusted blindly.

Keywords: geolocation, exhaustive tabulation, active measurements.

1 Introduction

Location-aware applications have recently become more and more widespread. Examples of such applications comprise targeted advertising on web pages, displaying local events and regional weather, automatic selection of a language to first display content, restricted content delivery following regional policies, and authorization of transactions only when performed from pre-established locations. Each application may have a different requirement on the resolution of the location estimation. Nevertheless, as IP addresses are in general allocated in an arbitrary fashion, there is no strict relationship between an IP address and the physical location of the corresponding physical interface.

Database-driven geolocation usually consists of a database-engine (e.g. SQL/ MySQL) containing records for a range of IP addresses, which are called blocks or prefixes. When coupled with a script embedded in a website and upon a client access to the website being detected, a request can be sent instantly to the database. This request can be to check if the IP address has an exact or longest prefix match (LPM) with a corresponding geographic location and coordinate. Since there is no actual measurement

M. Claypool and S. Uhlig (Eds.): PAM 2008, LNCS 4979, pp. 11–20, 2008.
© Springer-Verlag Berlin Heidelberg 2008

involved but merely a simple lookup, the request can be served in a matter of milliseconds. The expected time for which a website should be fully loaded, without causing any nuisance, is in general within one second. Most commercial database providers offer highly optimized scripts as well as abundantly documented application programming interfaces, which meet this short expected response time. The database-driven geolocation thus seems to be a useful approach.

Examples of geolocation databases are *GeoURL* [1], the *Net World Map* project [2], and free [3] or commercial tools [4, 5, 6, 7, 8, 9]. Exhaustive tabulation is difficult to manage and to keep updated, and the accuracy of the locations is unclear. In practice however, most location-aware applications seem to get a sufficiently good geographic resolution for their purposes.

In this paper, we try to better understand the resolution of geolocation databases, by comparing their location estimates with a well-known active measurements-based geolocation technique, CBG [10]. We show that, as expected, the geographic resolution of databases is far coarser than the resolution provided by active measurements, typically several times coarser than the confidence given by active measurements. As most geolocation databases do not give confidence in the accuracy of their location records, they are likely not to be trustworthy sources of geolocation information if precise IP address-level locations are required. Applications that require as much accuracy as possible would thus typically have to rely on active measurements, not databases. To improve the quality of current geolocation databases, we believe that the database records should contain information about the expected confidence in the location estimates.

The remainder of the paper is structured as follows. Section 2 introduces the datasets used. Section 3 studies the geographic resolution of databases. Section 4 describes our active measurements for geolocating Internet hosts. In Section 5, we compare the resolution of active measurements with location estimates from databases. Finally, we conclude in Section 6.

2   Datasets

During the past few years, a growing number of companies have spent a lot of effort in creating databases for geolocation purposes. Most of these companies, like Maxmind [11], Hexasoft [8] and Quova [9], provide commercially available databases with periodic updates. There are also freely available databases such as Host IP [3].

One of the problems of geolocation databases is that typically one does not know much about the methodology used by the database provider to gather their geographic information. One has to blindly rely on the claimed geographic resolution they provide. There are four basic geographic resolution levels that occur in most databases: zipcode, city, country and continent. Note that some databases may use more resolutions than those four, like regions that may relate to countries, continents, or some intermediate resolution. In most instances, we expect that the zipcode and the city granularity will be very similar. The country resolution is widely recognized to be the typical one that is reliable from databases. Many databases do not give any information about the expected geographic resolution of the database records, and when they do, not all records do contain this information. The price of commercial databases increases with improved geographic resolution, or with additional information about attributes of IP blocks like ISP, connection type of hosts, and in a single instance confidence about the location estimates. Note that we know one example of geolocation database that provides a notion of confidence related to the uncertainty about where the end-user actually lies compared to the location estimate [9]. This notion of confidence is however not quantitative, i.e. it does not express how far an IP address belonging to the IP block is expected to be from the location estimate provided, rather the type of host or connection that the host is using.

In the sequel of this paper, we restrict our attention to two databases. These commercial databases, GeoIP by Maxmind [11] and IP2Location by Hexasoft [8], are used because of their popularity (see [11, 8] for a listing of some of their customers) and their expected reliability. The number of IP blocks and the coverage in IP addresses of

Table 1. Overview of the 2 selected databases

Database	Public blocks	Special blocks	Total blocks	Public addresses	Total addresses
Maxmind	3,278,391	2	3,278,393	2,322,257,277	2,355,811,965
Hexasoft	5,111,309	44	5,111,353	3,991,797,760	4,294,967,296

the two databases is shown in Table 1. Maxmind contains more than 3 million blocks, and Hexasoft more than 5 million blocks. Note that a few blocks, called special blocks according to RFC3330 [12], should not be considered.

3   Geographic Resolution of Databases

Based on the information provided in the geolocation databases, it is hard to say anything about the actual geographic resolution of the location estimates. We merely know that most records contain either a city or a country name. 73.1% of the databases records in Maxmind contain a city name (66.6% for Hexasoft), then if no city name can be found, 3.4% of the records contain a country name (33.2% for Hexasoft). When neither a city name nor a country name is present in the record, a continent name or a federation of countries will typically be found. Note that sometimes records contain geographic coordinates only. While the area of countries and continents are well-defined, the area of a city depends much on what is meant by the boundaries of the considered city. For example, taking the largest 250 cities in the world[1] shows well how much the area of a city can vary, especially depending on whether the suburbs or the "metro" area are considered to be part of the city or not.

When we analyze the number of unique cities in both Maxmind and Hexasoft, we obtain 110, 349 unique cities in Maxmind and 15, 133 in Hexasoft. 100, 087 cities in Maxmind occur each in a single IP block (12, 918 for Hexasoft), and 10, 262 cities occur each in multiple IP blocks (2, 215 in Hexasoft). When several IP blocks have the same city information, they will have the same location estimate in the database. Note that a city is defined by a city name, but also a country and a continent when this

---

[1] http://www.citymayors.com/statistics/largest-cities-area-250.html

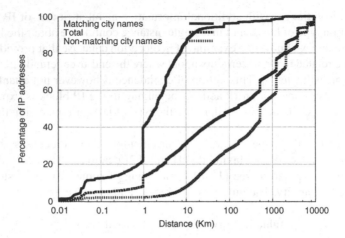

Fig. 1. Difference in location between Maxmind and Hexasoft

information is available in the databases. Some city names occur in several countries and/or continents. When we compare the occurrence of unique city names (string-wise), we observe that among a total of $7,844$ unique city names present in the databases, $7,618$ are present in one database only, and 226 are in both.

In geolocation databases, a unique location is associated to a given city. It is thus impossible to infer directly the geographic resolution used by the databases by comparing the location estimates of different IP blocks for a given city. However, we can compare the location estimates from Maxmind and Hexasoft, hoping that the difference between their location estimates will give us an indication of their geographic resolution. We rely on a free database, Host IP [3], that contains $1,356,506$ IP blocks, to perform lookups in the two other databases. For each IP block of Host IP, we take an IP address and use it to lookup the two databases. We then compute the difference between the two location estimates returned by the databases. Figure 1 displays the cumulative distribution of the distance between the locations given by the two databases when performing a lookup on IP addresses from the Host IP database. We provide three different curves, one for the distribution of the distance when the city strings match between Maxmind and Hexasoft, when they do not match, and irrespective of the city-level match. Among the $1,264,892$ IP addresses looked up, $377,736$ have the same city-level name in the databases, while $887,156$ do not have matching city names. We see on the curve that corresponds to matching cities that the difference in location between the databases tends to be far smaller than when the city names do not match. Depending on whether the city names match between the two databases entries, the typical distance between their location estimates differs much. When the IP blocks from the two databases have the same city name information, their locations are very close, typically less than 10Km. When the city names do not match on the other hand, the locations differ more than usual. Globally, about 50% of the IP lookups give a difference smaller than 100Km. If the differences observed between the databases were to reflect in some way differences in geographic resolutions used by them, then we would deduce that those resolutions go from 1Km up to thousands of Km.

4   Measurements-Based Geolocation

Given that we cannot obtain the actual geographic location of many IP addresses in the Internet, we need to rely on location estimates. To obtain location estimates for a large enough number of IP hosts, we need accurate location estimates. For this, we rely on active measurements. Active measurements have the advantage of providing an explicit estimate of their accuracy.

Previous works on measurement-based geolocation of Internet hosts [13, 14] use the positions of reference hosts, called landmarks, with a well-known geographic location as the possible location estimates for the target host. This leads to a discrete space of answers; the number of answers is equal to the number of reference hosts, which can limit the accuracy of the resulting location estimation. This is because the closest reference host may still be far from the target. To overcome this limitation, the authors of [10] propose the Constraint-Based Geolocation (CBG) approach, which infers the geographic location of Internet hosts using *multilateration*. Multilateration refers to the process of estimating a position using a sufficient number of distances to some fixed points. As a result, multilateration establishes a continuous space of answers instead of a discrete one. This multilateration with distance constraints provides an overestimation of the distance from each landmark to the target host to be located, thus determining a region, *i.e.* confidence region, that hopefully encloses the location of the target hosts [10]. For instance, the confidence region allows a location-aware application to assess whether the estimate is sufficiently accurate for its needs.

Although showing relatively accurate results in most cases, these measurement-based approaches may have their accuracy disturbed by many sources of distortion that affect delay measurements. For example, delay distortion may be introduced by the circuitous Internet paths that tend to unnecessarily inflate the end-to-end delay [15, 16, 17] and by the potential existence of bottleneck links along the paths. To deal with these sources of distortion, *GeoBuD, Octant*, and *TBG* were proposed by [18, 19, 20]. The GeoBuD technique shows that estimating buffering delays, by *traceroute* measurements, at intermediate hops along the traceroute path between a landmarks and a target host enables to improve the accuracy of geolocation of Internet hosts. In the same way, Topology-Based Geolocation (TBG) and Octant which are an extension of multilateration techniques with topology information were proposed. TBG additionally uses inter-router latencies on the landmark to target network paths to find a physical placement of the routers and target that minimizes inconsistencies with the network latencies. TBG relies on a global optimization that minimizes average position error for the routers and target. Octant differs from TBG by providing a geometric solution technique rather than one based on global optimization. Although it considers intermediate routers as additional landmarks, Octant also uses geographic and demographic information. Geographic and demographic constraints are used in Octant to reduce the region size where the target may be located. Only landmasses and areas with non-zero population are considered as possible target locations [19]. Furthermore, it takes into account queuing delays by using height as an extra dimension. It requires significantly computational time and resources. All these techniques generate a huge amount of overhead in the network for a small gain in accuracy.

To illustrate the marginal improvement of complex measurement-based geolocation techniques, we do not only consider CBG, but also add to it estimation of the bottleneck bandwidth on the path. The bottleneck bandwidth can be defined as the maximum throughput that is ideally obtained across the slowest link over a network path. CBG with bandwidth estimation allows the improvement of the geolocation estimation given by CBG. Additional delay distortions caused by the bottleneck along the path are removed from the overestimations of distance constraints that define the region enclosing the target host in CBG, allowing tighter overestimations that result in a smaller region. Smaller regions that still enclose the target host provide a more accurate location estimation.

4.1 CBG with Bandwidth Estimation

To estimate the bottleneck bandwidth over a network path between each landmark and a given target host, we use *SProbe* [21]. SProbe estimates bottleneck bandwidth in uncooperative environments, *i.e.* a measurement software is only deployed locally on the measurement host. SProbe relies on the exploitation of the *TCP* protocol. It sends two *SYN* packets to an inactive port on the remote host to which it appends 1460 bytes of data. Since the port is inactive, the remote host answers to these packets with two *RST* packets of 40 bytes each. For the native traceroute used by Octant, TBG, and GeoBuD, three packets are sent to each intermediate hops between a source and a destination causing an important overhead. SProbe produces accurate and fast estimates using little amount of probing data, so that it can scale to a large number of estimates.

For our evaluation, we rely on 39 *PlanetLab* nodes [22] as landmarks and we use a subset of the two commercial databases (Maxmind and Hexasoft) as input for hosts to be localized. Each landmark estimates the bottleneck bandwidth towards a given target host by sending 7 SYN packets. We found in Section 3 that there are 226 city names that are unique and can be found in both databases. Using these city names we find 41, 797 IP blocks from Maxmind matching those city names. Since we need "pingable" addresses within each IP block to be used in measurements, we use the single ping approach to find at least one IP address per block. The single ping approach consists in brute-force probing all IPs within a prefix, and stopping the probing within the prefix as soon as a single IP address has answered. We find 18, 805 IP blocks which have at least one pingable IP address for Maxmind. For the Hexasoft database, we have 41, 758 IP blocks among which 15, 823 contain at least one pingable IP address. Using the set of pingable addresses, Figure 2 presents the cumulative distribution of the confidence region in km^2 for location estimates in both the Maxmind and Hexasoft databases. Figure 2(a) shows that CBG with bandwidth estimation assigns a confidence region with a total less than 10^4 km^2 for about 20% of the location estimates, whereas the basic CBG has only 10% for the same confidence region. For IP addresses that are given a confidence region between 10^4 km^2 and 10^6 km^2, bandwidth estimation is less and and less useful. Finally, when the confidence region is larger than 10^6 km^2, bandwidth estimation is useless, or even makes the confidence region larger than the classical CBG technique.

Measurement-based geolocation techniques assume that the target host is able to answer measurements. Active measurements will be impractical when we rely on *ICMP*

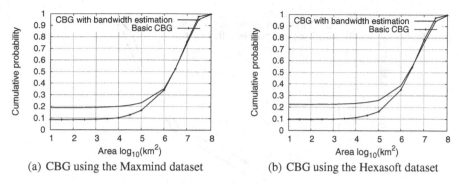

(a) CBG using the Maxmind dataset   (b) CBG using the Hexasoft dataset

Fig. 2. Confidence region

echo probes for instance, which can be filtered by a firewall. We observe that for most IP blocks, we get only a few IP addresses that answer our probes, typically only one.

5   Comparison Between Databases and Active Measurements

Having discussed the geographic resolution of geolocation databases in Section 3 and presented the confidence area obtained with active measurements in Section 4, we use the active measurements introduced in Section 4 to check the resolution of geolocation databases. When comparing geolocation based on active measurements and databases, several situations may occur. One possibility is when databases and active measurements give the same location for an IP address, i.e. databases give a location that lies within the confidence region given by active measurements. This situation is not typical, given the coarse geographic resolution of database records. When location estimates from the databases do not belong to the confidence region provided by active measurements, we would tend to doubt the accuracy of databases rather than expecting that the confidence region suffers from measurements biases, as the confidence region is made from higher bounds on the distance constraints.

Let us now measure the distance between the border of the confidence region given by CBG and the location estimates of the databases. If CBG is correct in its estimation of the location, then this distance should provide a lower bound on the actual geolocation error made by the database. Figure 3 shows the cumulative distribution of the minimal distance between the location estimates of the Maxmind dataset (results for Hexasoft are similar) and the border of the confidence region given by CBG, with and without using bandwidth estimation. This minimal distance first tells whether the location estimates from databases are within the confidence region or not. If the distance is negative on Figure 3, it means that databases are within the confidence region. If the confidence region is small and the location estimate of the database lies within the confidence region, then we expect that it is likely that the database estimate is correct. We observe on Figure 3 that more than 90% of the probed IP addresses have a database location estimate that lies outside the confidence region, and quite far away from it. Note that in a few cases the distance on Figure 3 is negative and large, meaning that the confidence region is pretty large.

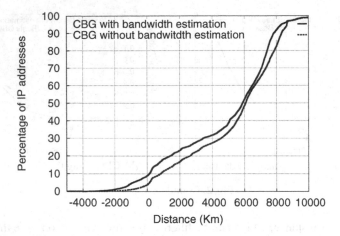

Fig. 3. Distance between the database results and the border of the CBG confidence region (Maxmind dataset)

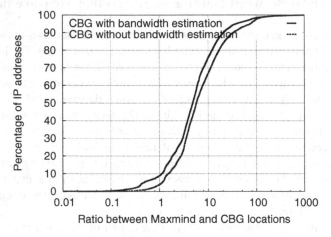

Fig. 4. Ratio of the distance of the databases to the distance of CBG with respect to the CBG location estimate (Maxmind dataset)

The large distances shown in Figure 3 suggest that the geographic resolution of databases is poor, compared to the confidence region given by CBG. To quantify the relative resolution of databases compared to the confidence region given by CBG, we plot in Figure 4 the ratio of the difference between the CBG estimate and the locations given by the Maxmind dataset (results for Hexasoft are similar), divided by the uncertainty in the CBG estimate (radius of the confidence region). Let us denote the location given by CBG by $loc_{cbg}(IP)$, the radius of CBG's confidence region by $radius_{cbg}(IP)$, and the location given by a database by $loc_{database}(IP)$, then the ratio we compute is $\left| \frac{loc_{database}(IP) - loc_{cbg}(IP)}{radius_{cbg}(IP)} \right|$. A ratio smaller than 1 means that the location estimate given by the database is within the confidence region. In this case, we would tend to trust the location estimate given by the database. A ratio larger than 1 means that the location estimate given by the database lies outside the confidence region. In that case, it is likely that the geographic resolution of the database is too coarse to give an accurate location estimate for the considered IP address. We observe on Figure 4 that the ratio is typically far larger than 1, meaning that the geographic resolution of the databases compared to the confidence in the active measurements estimates is poor, relative to the confidence region of CBG. For only less than 10% of the probed IP addresses, the databases have a good enough geographic resolution to make them comparable to the accuracy of active measurements. Note that those results do not suggest that location estimates provided by databases are incorrect, but rather that the geographic resolution at which databases give mappings from IP blocks to locations are too coarse to provide accuracy at the level of individual IP addresses.

6 Conclusion

In this paper, we assessed the geographic resolution of geolocation databases. We described the typical content of such databases, showing that they do not contain information to give confidence in the expected accuracy of their location estimates. We illustrated the relative coarse resolution databases provide, by showing how large the span of cities is, and how much the location estimates differ between the considered databases.

We carried out active measurements in order to compare the geographic resolution of databases to a more accurate standard. We quantified the accuracy of active measurements, and tried to improve them by adding bandwidth measurements to reduce the bias from bottleneck links.

Our comparison of the active measurements and the location estimates from the databases demonstrated the coarse geographic resolution of databases location estimates. We showed that not only the distance between the location estimate of the databases and the location given by active measurements is very large, but that also difference between the database location estimates from the active measurements estimates, divided by the accuracy expected from the active measurements, is very large.

Our work shows that the geographic resolution of geolocation databases is coarse compared to the one of active measurements. That does not mean that the location estimates given by databases are not good enough. Information about the geographic resolution of the databases can be embedded in them, for example by giving an estimate of the city-level span for each record. In general, we do not expect that active measurements will be so helpful to improve the geographic resolution of geolocation databases, simply because databases work at the level of IP blocks. However, in particular cases where better accuracy is required for specific IP addresses, active measurements have great potential to provide better location estimates than databases.

Acknowledgments

Bamba Gueye is supported by the IST ANA project.

References

1. GeoURL, http://www.geourl.org/
2. Net World Map, http://www.networldmap.com/
3. Host ip, http://www.hostip.info/
4. Digital Island Inc, http://www.digitalisland.com/
5. Akamai Inc, http://www.akamai.com/
6. GeoNetMap, http://www.geobytes.com/GeoNetMap.htm
7. WhereIsIP, http://www.jufsoft.com/whereisip/
8. Ip2location, Hexasoft Development Sdn. Bhd, http://www.ip2location.com/
9. GeoPoint, http://www.quova.com/
10. Gueye, B., Ziviani, A., Crovella, M., Fdida, S.: Constraint-based geolocation of internet hosts. IEEE/ACM Trans. Netw. 14(6), 1219–1232 (2006)
11. MaxMind LLC, MaxMind, http://www.maxmind.com
12. IANA, Special-use IPv4 addresses, Tech. Rep., Internet RFC 3330 (September 2002), http://www.rfc-editor.org/rfc/rfc3330.txt
13. Ziviani, A., Fdida, S., de Rezende, J.F., Duarte, O.C.M.B.: Improving the accuracy of measurement-based geographic location of Internet hosts. Computer Networks 47(4), 503–523 (2005)
14. Padmanabhan, V.N., Subramanian, L.: An investigation of geographic mapping techniques for Internet hosts. In: Proc. of ACM SIGCOMM, San Diego, CA, USA (August 2001)
15. Tangmunarunkit, H., Govindan, R., Shenker, S., Estrin, D.: The impact of routing policy on internet paths. In: Proc. of IEEE INFOCOM, Anchorage, AK, USA (April 2001)
16. Subramanian, L., Padmanabhan, V., Katz, R.: Geographic properties of Internet routing. In: Proc. USENIX, Monterey, CA, USA (June 2002)
17. Zheng, H., Lua, E.K., Pias, M., Griffin, T.: Internet Routing Policies and Round-Trip-Times. In: Proc. of PAM Workshop, Boston, MA, USA (April 2005)
18. Gueye, B., Uhlig, S., Ziviani, A., Fdida, S.: Leveraging buffering delay estimation for geolocation of Internet hosts. In: Proc. IFIP Networking Conference, Coimbra, Portugal (May 2006)
19. Wong, B., Stoyanov, I., Gün Sirer, E.: Geolocalization on the internet through constraint satisfaction. In: Proceedings of the 3rd conference on USENIX Workshop on Real, Large Distributed Systems
20. Katz-Bassett, E., John, J.P., Krishnamurthy, A., Wetherall, D., Anderson, T., Chawathe, Y.: Towards ip geolocation using delay and topology measurements. In: Proc. of ACM SIGCOMM Internet Measurement Conference, Rio de Janeiro, Brazil (October 2006)
21. Saroiu, S., Gummadi, P.K., Gribble, S.D.: Sprobe: A fast technique for measuring bottleneck bandwidth in uncooperative environments. In: Proc. of IEEE INFOCOM, New York, NY, USA (June 2002)
22. PlanetLab: An open platform for developing, deploying, and accessing planetary-scale services (2002), http://www.planet-lab.org

Observations of IPv6 Addresses

David Malone

Hamilton Institute, NUI Maynooth*
David.Malone@nuim.ie

Abstract. IPv6 addresses are longer than IPv4 addresses, and are so capable of greater expression. Given an IPv6 address, conventions and standards allow us to draw conclusions about how IPv6 is being used on the node with that address. We show a technique for analysing IPv6 addresses and apply it to a number of datasets. The datasets include addresses seen at a busy mirror server, at an IPv6-enabled TLD DNS server and when running traceroute across the production IPv6 network. The technique quantifies differences in these datasets that we intuitively expect, and shows that IPv6 is being used in different ways by different groups.

1 Introduction

IPv6 uses an address space that is far larger than could be consumed by devices in the near future. The reason for such a large address space is to try to make address management easier, both when numbering hosts within subnets and when numbering networks within the Internet. The hope is that addresses and subnets can be assigned according to logical schemes rather than assigning addresses in the most compact way.

This is in contrast to allocation of addresses for IPv4; in the CIDR world it is unclear where the network address ends and the host address begins. Some deductions can be made about IPv4 addresses: we can consult IANA and whois databases to determine who addresses have been assigned to. However, unless whois/DNS information is well maintained, it will be difficult to know how addresses are actually being used.

In this paper, we show how to use IPv6's extra expressiveness to infer things about how addresses are being used. As with IPv4, we can consult IANA in order to discover which registrar has been assigned the address. However, we can also identify people who connect to the IPv6 Internet using mechanisms such as 6to4 [5] and Teredo [7]. As there are standard procedures for allocating host IDs, we can identify auto-configured hosts and other address schemes in use.

Such an analysis of IPv6 address is not difficult in itself; a competent IPv6 network engineer could perform this analysis by glancing at an address. However, we will automate this analysis and apply it to large datasets. The first dataset is the IPv6 addresses observed in the wild at ftp.heanet.ie. Our second is the set of recursive DNS servers making queries to ns6.iedr.ie, an authoritative server for the *ie* domain. Our third is based on addresses responding to a traceroute through the IPv6 routing infrastructure. We aim to see what can be learned about IPv6's deployment in each situation through the observation of live addresses. Despite the limited nature of the datasets, we see interesting variations between them.

---

* I would like to thank HEAnet and the IEDR for providing access to their logfiles.

M. Claypool and S. Uhlig (Eds.): PAM 2008, LNCS 4979, pp. 21–30, 2008.

This is not the only study that presents techniques to assess the state of the IPv6 Internet, but we believe this is the first to focus on addresses as the primary source of information. For example, [3,11] analyse traffic seen at public 6to4 relays, considering indicators such as traffic levels, ports used and numbers of 6to4 clients. [1] describes a repository of traffic data, including IPv6 traffic, but also aims to anonymise the traffic, in the process scrambling much of the data that we aim to analyse. Work such as [8] focuses on routing tables and allocation of address blocks, but this exposes no information beyond the BGP prefix. Others have used active probing to measure IPv6 topology [4] or compare performance to IPv4 [2], focusing on the connectivity graph or performance of the network, rather than configuration details. Operators have also reported breakdowns of traffic volumes by application, using traditional indicators such as ports.

2 Address Analysis Technique

2.1 Prefix Analysis

The breakdown of IPv6 address space is described by several IANA registries. The overall breakdown of address space is described in the `ipv6-address-space` registry. Smaller chunks are then described in more detail by other registries. In the case of other global addresses, we can use the `ipv6-unicast-address-assignments` to identify the RIR that addresses have been assigned to.

This analysis is the same sort of classification that can be performed on IPv4 addresses. However, in some cases an IPv6 address will provide details about how IPv6 has been deployed. In particular, we can identify users of 6to4 (`2002::/16`), Teredo (`2001::/32`, formerly `3ffe:831f::/32`) and 6bone allocations (`3ffe::/16`, formerly `5f00::/8`).

2.2 Host ID Analysis

Just as the prefix can tell us where an address may be (al)located or if certain transition techniques are in use, the host-id can also give us information about how IPv6 is configured on a node. This sort of information is unavailable in IPv4 or in IPv6 if studies are solely based on address block or routing table information.

Perhaps the most common mechanisms for assigning host IDs are manual configuration (on routers and some servers) and autoconfiguration (based on the MAC address of a device). Autoconfiguration can usually be identified because of the way that MAC address are converted to host IDs. In particular, the dominance of EUI-64 based addressing and Ethernet/WiFi cards with vendor-assigned addresses leads us to expect `fffe` as the middle 16 bits of the host ID and the 7th bit of the host ID will be set.

ISATAP [12] is a technique that uses IPv4 as a layer 2 for IPv6, and it has a technique for generating an IPv6 host ID from the underlying IPv4 address. The construction leads us to expect the first quad (16 bits) of the host ID to be `0000` or `0200` and the second quad to be `5efe`, and these can be used as a test for ISATAP addresses.

ISATAP is not the only scheme that uses IPv4 addresses as a way to generate host IDs. IPv6 address parsers will usually allow the last two hex quads of the host ID to be

```
00ad 00ba 00be 00d0 00da 00ed 0ace 0ada 0add 0ade 0b00 0b0a 0b0b 0baa 0bad 0bea 0bed 0bee 0c00 0c0b 0c0d 0cab 0d0b
0d0c 0d0d 0d0e 0dab 0dad 0deb 0dee 0ebb 0f00 0f0b 0f0d 0f0e 0fad 0fae 0fed 0fee abba b00b b0b0 b0de baba babe bade
baff bead beef c0c0 c0ca c0d0 c0da c0de c0ed c0ff cafe cede d00b d0d0 d0de dada dead deaf deed f00d f0ad face fade
faff feed 1337 0000 1111 2222 3333 4444 5555 6666 7777 8888 9999 aaaa bbbb cccc dddd eeee ffff 00ff abab
```

Fig. 1. Hex words users might use in IPv6 addresses

written as a traditional IPv4 dotted decimal. We use the following heuristic to identify host IDs that have the last two quads generated from a IPv4 address: an address is v4-based if (a) it is a 6to4 address and the second quad is the same as the seventh quad; or (b) the fifth and sixth quads are zero, the seventh and eighth quads are different, and the resulting IPv4 address would not be in IANA reserved/multicast address space.

We have based this heuristic on several factors. One is that 6over4 uses an IPv4 address padded with zeros and, as noted in [11], some 6to4 implementations do the same. Also, some sites derive their manually configured IPv6 address scheme from their IPv4 scheme. This test has weaknesses; we will discuss its effectiveness in Section 4.

Teredo also uses a special host ID based on two IPv4 addresses (the address of the NAT box and the address of the Teredo server). Since Teredo uses an easily identifiable prefix, we identify such host IDs based on the prefix.

In IPv4 networks, it is common for addresses to be assigned dynamically by DHCP from a pool. This currently seems less common in IPv6 networks, maybe because of the wide availability of autoconfiguration and relatively slow development of DHCPv6-capable servers. Concerns were raised because a fixed host ID, generated from a MAC address, would allow the tracking of devices as they moved from network to network. In response to this, a technique for randomly generating IPv6 host IDs was specified [10], which is now available in many IPv6 implementations. This *privacy addressing* uses a cryptographic hash to generate the host ID, and then clears the 6th bit.

This provides us with a technique to identify privacy addresses. A cryptographic hash should produce 0 and 1 bits in equal proportions. For a 63-bit output the Law of Large Numbers says that the majority of privacy addresses will have around 32 bits set. The actual technique used to identify privacy addresses is to first determine if the address can be identified as some other sort of address, and if not it is considered as a candidate privacy address. The address must then satisfy the following: the host ID must have the 6th bit clear; the host ID must have between 27 and 35 set bits; the first 32 bits must have between 9 and 21 set bits; the last 32 bits must have between 10 and 22 set bits; the host ID must not have two or more 'words' in it (as shown in Fig. 1).

These criteria are designed to cover the majority of privacy addresses, while rejecting patterns that are likely to have been manually configured, such as ::ffff:ffff and feed:babe:dead:beef. We can calculate the proportion of random addresses that satisfy these conditions on the number of set bits as

$$\frac{1}{2^{63}} \sum_{\substack{9 \leq i \leq 21, 10 \leq j \leq 22 \\ 27 \leq i+j \leq 35}} \binom{31}{i}\binom{32}{j} \approx 0.7335. \tag{1}$$

Correcting for privacy addresses that are identified as being in some other type results in a insignificant change in this fraction. Thus, we expect this test to identify about three quarters of all privacy addresses.

The main type of host ID that we have not considered is manually-assigned host IDs. We cannot hope to identify all host IDs that are assigned directly by humans (or their scripts). However, humans are likely to opt for simple addresses that are easy to remember. One class of these are addresses ending in something simple, such as : : 1 or : : 53. We attempt to identify these as addresses with the first 56 bits being 0, and call them *low* addresses. A class of address that humans are likely to be drawn to is those with regular patterns or words. Host IDs composed of 4 quads from Fig. 1 are *wordy*.

When attempting to identify a host ID, we take the first matching test from the following list: ISATAP, Teredo, autoconf, low, IPv4-based, wordy, privacy.

3   Data Sets

3.1   HEAnet Mirror Server

This dataset is based on the log files from the Apache server running on ftp.heanet.ie, a mirror server located in HEAnet, Ireland's research and education network. HEAnet's mirror server began offering IPv6 services publicly around May 2002, when a AAAA DNS record for the server was added. It mirrors a large number of projects, including Sourceforge, various Linux distributions, Apache, PuTTY, Mozilla, etc. It has a large user population and is the twentieth most visited site hosted in Ireland, according to Netcraft. There is no specific IPv6-related content on the site, though the software available may attract technically curious users.

Load on mirror servers can be highly variable. Peaks can be caused by new software releases or changes in available mirrors. For example, for a period ftp.heanet.ie was the only continuously operational Sourceforge mirror, resulting in increased load because of the sticky cookie used to select a Sourceforge mirror.

The dataset begins on 7 December 2003 and ends on 3 August 2007, over 1300 days. On some days during the period no data is available because of maintenance, service interruptions or log files no longer being available. One substantial gap is from mid-August 2005 to the end of 2005, due to an absence of log files.

From the beginning of the data, we have a list of the time and address of each IPv6 request to the server and summary IPv4 statistics. From 1 February 2005 onwards, Apache logs in the combined log file format entries for both IPv4 and IPv6 accesses are available. Fig. 2(a) shows the number of IPv6 hits (i.e. individual HTTP requests) on all days on which there were more than 1000 IPv6 requests. Daily IPv4 statistics are also shown where available, in some places interpolated from monthly statistics.

We aim to present statistics that account for, or make apparent, missing data and fluctuations in load. For example, to account for trends in IPv6 usage, we should factor out missing data or changes in load. One way to do this is to normalise by the IPv4 hits. For this to be valid, we need to know if IPv4 and IPv6 hits are correlated.

Fig. 2(b) shows a scatter plot of IPv4 vs IPv6 per-day hits. We plot points only where we have per-day statistics for both IPv4 and IPv6. The region excludes about 5 outlying points. It seems the majority of days have the IPv4 load between about 400 and 1200 times the IPv6 load. However, there are a considerable number of points with higher IPv6 load. This suggests one should be cautious about blindly normalising by IPv4 hits, though IPv4 and IPv6 load do seem correlated.

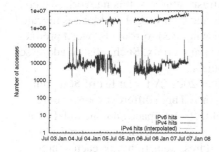
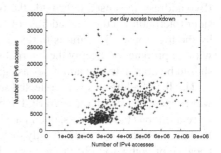

Fig. 2. HEAnet load statistics: (a) Per day hits (log scale), (b) scatter plot of IPv4 vs IPv6 hits

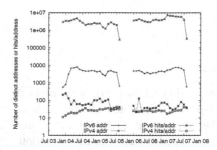

Fig. 3. Distinct addresses/hits per IP each month

We will be interested in the number of distinct addresses seen, as addresses are the fundamental unit of our analysis. Fig. 3 shows the number of distinct IPv6 addresses seen during each month. The number of distinct IPv4 addresses is also shown for comparison. We see that, except for the first few months, the pattern of fluctuations is similar for both IPv4 and IPv6, suggesting common causes for the fluctuations, such as those mentioned above. The dips in August 2005/2007 are caused by partial data for these months. Fig. 3 also shows the mean number of hits made per IP address. While these statistics were initially quite different, it seems as if they may be coming closer.

3.2   IE IPv6-Enabled Nameserver

The IE top-level domain is served by a number of IPv6-capable name servers. The IEDR, Ireland's domain registry, operate one of these, which has been advertised in the root zone since September 2004. Log files showing all queries to this name server were available for the dates 22 April 2007–20 May 2007. The log files record the date of the DNS requests and the IPv6 address making the request. The server only deals with IPv6 requests, so no comparable IPv4 statistics are available.

3.3   Traceroute Data

In this section we consider quite a different dataset. The global IPv6 routing table is still quite compact, with only around 1000 prefixes present. We can consider what sort of

IPv6 addressing is used to provide the routing infrastructure for this network. We consider tracrouting to the : : 1 address of each prefix and recording the addresses revealed by the traceroutes. The aim is to reveal the addressing used to route between prefixes, without probing too deeply the internal structure of any prefix. Such a list of addresses should be dominated by addresses assigned to routers.

The list of addresses was collected in September 2007. A target list of 866 prefixes was prepared based on the IPv6 BGP table at HEAnet. Three different source addresses were used: one 6to4 address, one in a commercial ISP's PA space and one in HEAnet's PA space. We expect to see slightly different lists of addresses for each source address, because of both variability in routes and source address selection. For each source address, a list of intermediate router addresses was produced using traceroute. The three different source addresses produced 1558, 1687 and 1698 addresses respectively.

4 Results

4.1 HEAnet Mirror Server

We analyse the data from Section 3.1 first. Fig. 4(a) shows the proportion of IPv6 addresses in the 6bone, global production, 6to4 and Teredo address ranges from month to month. We plot the number of addresses falling into each prefix each month divided by the total number of distinct addresses seen during that month. We do not show results for a small number of local addresses, such as the loopback and link-local addresses.

We see substantial activity in the global and 6to4 address space, with the fraction of global production addresses showing an increasing trend. As expected, 6bone addresses were on a gradual decline, until a sharp drop in May 2006 before their retirement in June 2006. HEAnet did not carry routes to 6bone addresses after 6/6/2006, so after this date access to ftp.heanet.ie was not possible from 6bone.

Initially, we see a handful of Teredo-based addresses. However, since mid-2006 there has been a substantial increase in the use of Teredo clients. While this growth took place at the same time as early Windows Vista deployment, the User-Agent information indicates a mix of operating systems, mainly Windows XP, Linux and FreeBSD.

Fig. 4(b) shows the proportion of global addresses seen that were allocated by each RIR. Roughly, RIPE covers Europe, ARIN covers north America, APNIC covers the

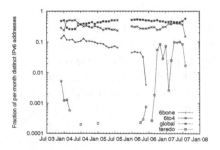
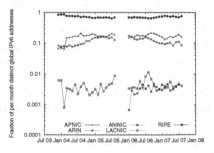

Fig. 4. Analysis of (a) all addresses by prefix, (b) global addresses by region

Asia/Pacific region, LACNIC covers Latin America and the Caribbean and AfriNIC covers Africa. As a mirror in Europe, we expect a majority of accesses to come from RIPE. Unsurprisingly, the statistics confirm this.

Accesses from outside Europe are more interesting. These users have no particular reason to select a mirror in Ireland and so may give some indication of relative levels of activity. Initially, activity from ARIN and APNIC are at a similar level. Accesses from APNIC regions jump sharply in March 2004 and then slowly-increase. Accesses from ARIN gradually increase over time, catching up on the APNIC around February 2005, and then slowly decline. We cannot be certain if this is a change in the overall IPv6 node population in these areas, or particular content causing differential activity between regions. We see activity from younger registries at a low but consistent level.

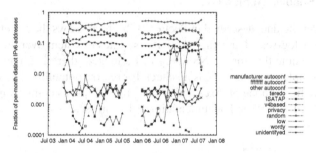

Fig. 5. Analysis of Host IDs

Now we turn to the host ID. Fig. 5 shows how host IDs have evolved over time. Consistently there are about 10% of addresses that we cannot classify. The dominant technique is autoconfiguration based on vendor-assigned MAC addresses. The next most common technique seems to be IPv4-based addresses. Most (95%) of these addresses are 6to4 addresses using an IPv4-based host ID. Manually examining the remaining 5% of those identified as IPv4-based shows false positives, but the majority of results look correct. A substantial number of addresses allocated to BTexact's Tunnel Broker are identified as being IPv4-based: these also seem likely to be correct.

The next most common host IDs are addresses with only the low byte non-zero. These addresses do not seem to show evidence of any particular technology dominating, maybe indicating a mix of manual configuration and scripting. The wordy addresses are a smaller proportion of overall host IDs than the low addresses, though both types are substantially more common than would be expected at random.

About 4% of addresses are identified as privacy addresses. From Equation 1 we know our test under-reports, so the actual figure should be $4/.73$ or about 5.5%. Note, privacy addresses may, in a sense, be over-represented; while an autoconfigued address is fixed, a privacy address is periodically regenerated. We also show results for *random* addresses, which would have been classified as privacy addresses except the 6th bit was set. We see a tiny number of these *random* addresses. This indicates that the classification of address as as a privacy address is unlikely to include many false positives. When we inspect the random addresses, we see mostly random patterns and a few regular patterns that have been incorrectly matched.

Some addresses look autoconfigured but do not have the global bit set to a value we expected based on vendor-assigned MAC address. These may be generated from manually-assigned MAC addresses, may be soft MAC addresses on virtual machines, or may be manually-assigned host IDs. The data showed no addresses generated from the MAC range used by VRRP/CARP. We do find some addresses corresponding to autoconfiguration from the MAC broadcast address. This host ID may be the result of a failed Ethernet EPROM/faulty driver combined with autoconfiguration.

A small, but increasing, amount of ISATAP activity is visible. By comparing with Fig. 4(a) we see that by the end of our data the populations of 6to4, Teredo and ISATAP clients are roughly in a ratio of 30%:10%:0.3%.

4.2 IE IPv6-Enabled Nameserver

We now consider the data described in Section 3.2. Fig. 6 shows the results for the two months of data as log-scale bar charts. Activity is quite consistent over the two months. It is important to note that for a DNS query to be logged, there is no need for a TCP handshake to complete, and so there is no check for return routability. Thus, about 0.5% of queries come from 6bone prefixes and a handful of requests come from unassigned (2000:1::1), ULA ([6]) and documentation [9] addresses.

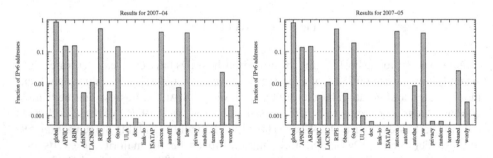

Fig. 6. Results for IEDR name server (log scale), April and May 2007

For comparison, Fig. 7 shows the corresponding results for ftp.heanet.ie. We expect a contrast between the client populations of these servers, because clients of mirror servers should not have much in common with recursive DNS servers. Indeed, when compared to ftp.heanet.ie, ns6.iedr.ie sees fewer requests from privacy, Teredo and ISATAP addresses. Even the proportion of 6to4 addresses is lower. The geographic distribution of the addresses also seems more even. We also see an increase in wordy and low addressing. This suggests that administrators shy away from transition mechanisms for recursive resolvers and opt for manually-assigned addresses.

4.3 Traceroute Data

Now consider the addresses observed in tracerouting across the IPv6 network from a commercial ISP, an NREN (HEAnet) and from a 6to4 network. Fig. 8 shows the

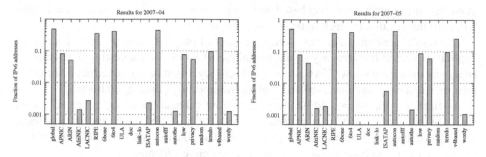

Fig. 7. Equivalent results for HEAnet ftp server (log scale), April and May 2007

breakdown of addresses observed from each of these sources. Note that the results are quite consistent with each other, but show differences when compared to Fig. 6 and Fig. 7. In particular, almost all addresses seen are global IPv6 addresses and most host IDs are either low or IPv4-based. Some "autoconf" addresses are observed, however this is a misnomer in the case of routers, as routers can use EUI-64 based addressing, but do not assign their own addresses based on IPv6 autoconfiguration. Again, no addresses generated from the VRRP MAC address were observed.

There is an absence of Teredo and ISATAP addresses, and 6to4 addressing is uncommon except where the probes are sent from a 6to4 source address. If the probe is sent from a 6to4 address, source address selection should cause the router to choose a 6to4 address for the response, if it has one. When compared to the results in Fig. 6 and Fig. 7 we see a more even distribution across all 5 RIRs, representing the indiscriminately global nature of the traceroute. There is still some systematic bias in favour of RIPE as all the source nodes were located in Europe, but this dataset shows the most even geographic distribution of addresses. Otherwise, both the ftp and DNS server see a broader spread of address types than traceroute does.

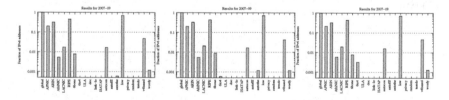

Fig. 8. Results for traceroute6 from Commercial ISP, HEAnet and 6to4, September 2007

5   Conclusion

We have presented a technique for analysing aspects of IPv6 addresses. We applied this technique to three different datasets. The results produced have shown consistency from month to month, suggesting that the technique should be stable enough to identify trends in IPv6 deployment. From the traceroute data we also see that the results are consistent across different networks. Comparing results between datasets, the technique quantifies differences we expect based on the operation of IPv6 networks. These results build confidence that the technique produces meaningful results.

The results are consistent with what we expect of the IPv6 Internet, but we have not explicitly verified the accuracy of our technique. It would be interesting to do this by analysing data from known IPv6 networks. Accuracy might be improved accounting for the times addresses are observed and spatially/temporally adjacent addresses.

As further work, we would like to look at how subnets are being allocated within organisations and provide a more detailed study of host IDs, including tracing them and linking them with manufacturers. We would also like to explore applications of this technique to log analysis and adaption of service delivery/content.

References

1. Cho, K., et al.: Traffic Data Repository at the WIDE project. In: USENIX FREENIX Track (2000)
2. Cho, K., et al.: Identifying IPv6 Network Problems in the Dual-Stack World. In: ACM SIG-COMM Workshop on Network Troubleshooting (2004)
3. Kei, Y., Yamazakim, K.: Traffic Analysis and Worldwide Operation of Open 6to4 Relays for IPv6 Deployment. In: Symposium on Applications and the Internet (SAINT) (2004)
4. CAIDA. Macroscopic IPv6 topology measurements (2006),
   http://www.caida.org/analysis/topology/macroscopic/IPv6
5. Carpenter, B., Moore, K.: Connection of IPv6 domains via IPv4 clouds. In: RFC 3056 (2001)
6. Hinden, R., Haberman, B.: Unique local IPv6 unicast addresses. In: RFC 4193 (2005)
7. Huitema, C.: Teredo: Tunneling IPv6 over UDP through NAT. In: RFC 4380 (2006)
8. Huston, G.: IPv6 BGP/CIDR reports, http://bgp.potaroo.net/index-v6.html
9. Huston, G., et al.: IPv6 address prefix reserved for documentation. In: RFC 3849 (2004)
10. Narten, T., Draves, R.: Privacy extensions for stateless address autoconfiguration in IPv6. In: RFC 3041 (2001)
11. Savola, P.: Observations of IPv6 traffic on a 6to4 relay. ACM SIGCOMM CCR 35 (2005)
12. Templin, F., et al.: Intra-site automatic tunnel addressing protocol (ISATAP). In: RFC 4214 (2005)

The New Web: Characterizing AJAX Traffic

Fabian Schneider, Sachin Agarwal, Tansu Alpcan, and Anja Feldmann

Deutsche Telekom Laboratories / Technische Universität Berlin
10587 Berlin, Germany
{fabian,anja}@net.t-labs.tu-berlin.de,
{sachin.agarwal,tansu.alpcan}@telekom.de

Abstract. The rapid advent of "Web 2.0" applications has unleashed new HTTP traffic patterns which differ from the conventional HTTP request-response model. In particular, asynchronous pre-fetching of data in order to provide a smooth web browsing experience and richer HTTP payloads (e.g., Javascript libraries) of Web 2.0 applications induce larger, heavier, and more bursty traffic on the underlying networks. We present a traffic study of several Web 2.0 applications including Google Maps, modern web-email, and social networking web sites, and compare their traffic characteristics with the ambient HTTP traffic. We highlight the key differences between Web 2.0 traffic and all HTTP traffic through statistical analysis. As such our work elucidates the changing face of one of the most popular application on the Internet: The World Wide Web.

1 Introduction

The World Wide Web [1] is one of the most popular applications of the Internet and runs primarily over the HTTP protocol. While HTTP (Hyper Text Transfer Protocol) [2] constitutes the session layer or messaging protocol of the Web, HTML (Hyper Text Markup Language) describes the content and allows authors of web content to connect up web pages through hypertext links or *hyperlinks*; an idea made popular by Tim Berners-Lee in the early 1990s and widely used today. In its classical form, users reach other pages or access new data by clicking on hyperlinks or submitting web based forms. In this basic HTTP request-response model each clicked link or submitted form results in downloading a new web page in response to the respective request.

The recent popularity of asynchronous communication enabled web sites has caused a significant shift from the classical HTTP request-response model of the Web. This asynchronous communication is commonly executed through AJAX (Asynchronous JavaScript and XML) [3], a compendium of technologies that enable web browsers to request data from the server asynchronously, i.e., without requiring human intervention such as clicking on a hyperlink or on a button. Consequently, HTTP requests are increasingly becoming automated rather than being human-generated. In this paper we use AJAX and "Web 2.0" interchangeably to refer to web applications that use this new paradigm on the Internet.

Contemporary web pages often contain embedded request-response functions comprising a JavaScript application engine that automatically executes in the background to asynchronously pre-fetch large quantities of data from the server. This intelligent

M. Claypool and S. Uhlig (Eds.): PAM 2008, LNCS 4979, pp. 31–40, 2008.

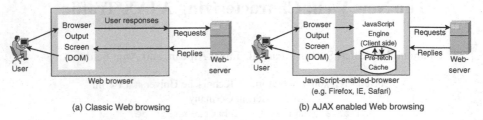

Fig. 1. Comparison of classical with AJAX enabled web applications

pre-fetching is often used to mask the round trip and transmission latency of Internet connections to give the user a 'smoother' web application experience. We highlight the differences in Figure 1. The JavaScript engine builds a local pre-fetched cache based on the user's interaction with the web application and executes parts of the application logic in the client's web browser itself instead of the web server. The prediction algorithms of any automated pre-fetching scheme usually results in significantly larger downloads as compared to user-initiated web browsing due to inaccurate guesses on part of the prediction algorithms about which data to pre-fetch. Even when the prediction is accurate, HTTP traffic inter-request-times are no longer lower-bounded by human response times (order of seconds) and may instead depend on the JavaScript code logic of the web application on the client machine.

Many popular web applications have adopted Web 2.0 technologies. One of the most popular and early adopter of AJAX is Google Maps. Its success encouraged the use of AJAX for building other interactive web applications. For example, many web-email offerings have transitioned into Web 2.0 applications in order to rival the look and feel of desktop email clients. Furthermore, some social networking web sites use AJAX technologies to offer rich and interactive user experiences. In this paper we explore the traffic characteristics of the most popular representatives of these AJAX based applications in our environment and contrast their characteristics to those of the overall HTTP traffic.

1.1   Related Work

A good overview of traditional Web protocols is given in the book by Krishnamurthy and Rexford [1]. One of the early works on characterizing the effect of HTTP traffic and HTTP pre-fetching is by Crovella [4]. It highlights the beneficial and unwanted effects of pre-fetching HTTP data, and hence further substantiates the importance of our analysis of Web 2.0 applications and their global effect on the Internet. There has been a vast literature on Internet web caching, e.g., [5,6,7]. However, the underlying motivation for using caching in all these studies has been on reducing the overall download latency of popular web sites and not facilitating low latency interactive Web 2.0 applications.

There are few studies focusing on the characteristics of AJAX-based traffic, although there exist several discussions, blogs and web sites about the end-user perceived latency of AJAX based applications (e.g., [8]). The novel aspect of our work is that we focus on the behavior of two large user populations and investigate multiple AJAX enabled applications.

1.2 Contributions

In this paper, we highlight the changing characteristics of Web traffic by comparing the traffic patterns of HTTP and Web 2.0 applications. For this we rely on several HTTP traces from large user populations in Munich, Germany and Berkeley, USA from which we extract popular AJAX application traffic.

From the statistical analysis of Web 2.0 traffic in comparison to all HTTP traffic extracted from the traces we show that the former's characteristics significantly differ from the latter's. Our work focuses on the number of transfered bytes, the number of HTTP requests issued and the times between subsequent request (inter-request-times). For example, Web 2.0 traffic has shorter inter-arrival-times due to the underlying human-independent automated data pre-fetching schemes.

Our work complements the efforts of the web developer community towards a better understanding of the Web 2.0 application characteristics. Some of our results may motivate the web developer community to design web applications that are friendlier to the underlying network, for example, by reducing the number of automated HTTP requests when possible.

The rest of the paper is organized as follows. We give a brief overview of the applications studied in this work and then describe our data collection process in Section 2. In Section 3 we present the results of our statistical analysis comparing AJAX traffic with the HTTP traffic. Finally, we conclude in Section 4.

2 Methodology

In order to determine which Web 2.0 applications to study we first examine the popularity of different applications (Section 2.1). Google Maps is among the most popular web applications and a nice example of a AJAX-enabled application. Therefore, we provide a high level overview of its communication patterns in Section 2.2. Finally, we detail how we extract application characteristics from our data sets in Section 2.3. Similar extraction methodologies (skipped for brevity) are used for the other AJAX applications.

2.1 Data Sets

We use packet level traces collected from two independent networks: the Münchener Wissenschaftsnetz (Munich Scientific Network, MWN) in Germany, and the Lawrence Berkeley National Laboratories (LBNL) in the USA. Both environments provide high speed Internet connections to their users. The MWN provides a 10 Gbps link capacity to roughly 55,000 hosts at two major universities and several research institutes, transferring 3-6 TB a day. LBNL utilizes a 1 Gbps upstream link, transferring about 1.5 TB a day for roughly 13,000 hosts. We base our analysis on three traces from network port 80 (the HTTP port). Two of these traces, MWN-05 and MWN-07, are from MWN while one trace, LBNL-07, is from LBNL. See Table 1 for information about the traces including: size, duration and start dates, total number of HTTP requests, and number of HTTP requests related to Google Maps.

We rely on packet level traces of large user populations as they provide the most detailed data. From these traces we reconstruct the HTTP request-response stream of

Table 1. Characteristics of the data sets

Trace	Start Date	Duration	Size	#Req Total	#Req GMaps
MWN-07	Feb 24th 2007	32h+	2.4 TB	30,0 M	222 K
LBNL-07	Mar 3rd 2007	~9h	214 GB	2,0 M	82 K
MWN-05	Oct 11th 2005	24h	2.5 TB	119 M	43 K

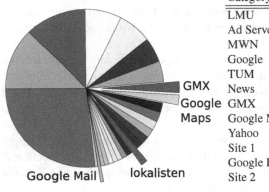

Category	Percent	Category	Percent
LMU	12.60%	Ebay	1.67%
Ad Server	12.26%	Site 3	1.45%
MWN	7.60%	lokalisten.de	1.40%
Google	6.90%	Bav. State	1.14%
TUM	5.05%	web.de	1.09%
News	4.23%	Site 4	1.01%
GMX	2.27%	Site 5	1.00%
Google Maps	2.04%	studivz.de	0.97%
Yahoo	1.97%	MSN	0.93%
Site 1	1.96%	Microsoft	0.76%
Google Earth	1.85%	Google Mail	0.65%
Site 2	1.81%	Youtube	0.55%
		other	26.83%

Fig. 2. Pie chart of the percentages of requests for the top 500 hostnames by categories. (Top 500 hostnames account for 53% of the total requests. Percentages are relative to the top 500.)

all connections. While one could use a variety of tools [1], we utilize the HTTP analyzer of Bro [10], a network intrusion detection system. Bro's policy script http.bro together with the policy scripts http-reply.bro and http-header.bro enable TCP stream re-assembly, basic HTTP analysis, and HTTP request-response analysis. We augmented the http-header.bro script to extract the times when the HTTP requests were issued. The resulting output file consisted of one-line summaries of each HTTP request containing (TCP) Connection ID, number of request in the connection, session ID, transferred bytes, three timestamps (request issued, cookie seen, request finished), requested hostname (servername[1]), prefix of the requested URL, and the HTTP status code for this request. Note that the number of transferred bytes does not include the HTTP header size. We only include requests for which we successfully record start and end times.

In order to determine the most popular AJAX enabled Web 2.0 applications we first identified the 500 most popular web servers[2] in the MWN-07 data set. We then grouped these into multiple categories for better visualization. The first set of categories contained the servers that are hosted by the two universities and the other research institutes (MWN). The next categories contained all request related to advertisements (Ad Server) and news web sites (News). Manual inspection showed that neither category contained

---

[1] We use server and host interchangeably in this discussion.

[2] Web server as specified by the hostname in the HTTP request.

many AJAX related requests. Some of the services offered by Google, including Google Maps and Google Mail use AJAX, while others like Google search, Google images and Google Earth, do not. Accordingly, we separate them into Google Maps, Google Mail, Google Earth, and all others (Google). Another popular Web-email service in Germany, that is also AJAX supported, is provided by GMX. Some categories include just a single popular site (Site 1, ..., Site 5), others are well known Web sites, e.g., Ebay and MSN. Figure 2 shows a pie chart of the number of requests per category for the MWN-07 data set. We find that GMX is the most popular AJAX based application with 2.27% of the requests followed by Google Maps which contributes 2.04%. Another AJAX-enabled social networking web site is lokalisten.de with 1.4%. Although Google Mail only accounts for 0.65% of the requests we include it as our fourth applications since this gives us two AJAX-enabled Mail applications by different providers. In terms of bytes the contributions are smaller, e.g., Google Maps with 1.41%. But all of the applications considered in this paper are among the top 500. We refer to these both most popular and AJAX-enabled applications as "Selected-4" in subsequent discussions.

2.2 Google Maps Communication

Google Maps is one of the first web applications to popularize AJAX technology. Consequently, it is widely considered as the canonical example of an AJAX application. AJAX uses the Document Object Model (DOM) [9] of the web browser such that it is no longer necessary to reload the entire web page each time it is updated. In this way it increases interactivity, speed, and usability.

Google Maps maintains multiple connections to different servers in the Internet that serve as back-ends for the Google Maps application. All connections use HTTP as the session protocol and take advantage of the advanced features of HTTP 1.1 [2] such as persistent HTTP connections for efficiency and pipelining for reducing latency, leading to multiple HTTP requests per TCP connection. In the context of Google Maps, most of these connections are used to fetch image tiles of the map. The others are used for control messages and for the initial transfer of the AJAX application (JavaScript code), the transfer of other GUI related pictures, and user queries. The connections carrying tile images can be identified by the servers they connect to.

2.3 Application Characterization Methodology

In this section we discuss how to extract application specific data from our data sets. For brevity reasons we focus on Google Maps traffic.

One of the challenges of identifying Google Maps traffic is that Google offers all its services on the same back-end server infrastructure (e.g., Google Maps, Google Search, Google Video, etc.) and uses a uniform key for all services. Therefore, the browser can reuse existing TCP connections to Google servers to issue Google search queries, image or video queries, as well as Google Maps queries. Separating Google Maps traffic from other Google services thus requires some effort. Moreover, to capture the user's interaction with Google Maps, we are not only interested in individual HTTP requests but also in the full set of HTTP requests within a Google Maps "session". Meaning all requests that are issued when a user connects to maps.google.com and then interacts with the application, e.g., by entering some location, by moving the map, or switching the zoom level. Accordingly, we group these requests to a Google Maps "session".

To identify Google Maps related requests among the very large number of HTTP requests within our traces we check if the hostname contains the string maps.google. To find the other requests by the same user we take advantage of Google's own session book-keeping mechanisms. Google uses cookies to mark all requests of a session by embedding a unique hash of its session ID[3]. We use this ID as our session ID as well and gather all other requests of this Google Maps session using the session ID. Unfortunately, there maybe additional requests to other Google services among the identified requests. We exclude these if they do not contain a Google Maps specific URL prefix. We found that /mt (map), /kh (satellite), /mld (route planning) and /mapstt (traffic) are related to the kind of map that is requested. /maps, /mapfiles and /intl are used for meta information. / and favicon.ico are not restricted to Google Maps use. A similar methodology is used for the other Selected-4 applications.

For comparison purposes, we also group requests of the complete HTTP traffic (ALL-HTTP), including requests of the Selected-4, into web sessions. In this case we cannot take advantage of cookies yielding session identifiers. Therefore we group those requests that come from the same client IP, go to the same server (IP) on the same server port. This aggregates connections from different client side ports.

For both Selected-4 sessions, and ALL-HTTP sessions we use a timeout[4] of 10 minutes. We compute per connection and per session statistics including number of transferred HTTP payload bytes, number of requests, their durations, and inter-request-times (IRT's) for the Selected-4 applications as well as ALL-HTTP traffic.

3   Characteristics of AJAX Traffic

In this section, we present the results of a statistical analysis of the characteristics of both ALL-HTTP and Selected-4 traffic. Almost all connections and sessions are usually comprised of multiple requests. However, we find significant differences in the session characteristics including: session life times, transferred bytes per session, number of requests within sessions, and inter-arrival-times of HTTP requests within sessions.

Most of the data is presented as probability density functions (PDF) although complementary cumulative distribution functions (CCDFs) are also shown. In order to capture the multiple orders of magnitude in the data we plot all CCDFs on a log-log scale and compute the PDFs of the logarithm of the data in order to be able to use a logarithmic X-axis. In addition, Table 2 presents mean and median values.

In our analysis we concentrate on the MWN-07 data set and only use the MWN-05 and LBNL-07 data sets to highlight some of the noticeable differences. Note that the 2005 data set was collected during Google Maps beta testing phase.

Figure 3 shows the CCDF of the number of bytes transferred in a single HTTP connection for ALL-HTTP and all Selected-4 applications for the MWN-07 data set. ALL-HTTP connections are clearly consistent with a heavy-tailed distribution over several

---

[3] The hash is located after the string PREF=ID= in the cookie.

[4] If the time between the end of a reply and the start of the next request is larger than 10 minutes a new session is started.

Table 2. Mean/Median Table for ALL-HTTP and Selected-4 applications in the MWN-07 data set. IRT's are Inter-Request(-arrival)-Times.

Application	#Requests	#Sessions		Bytes per Connection	Bytes per Session	#Req per Connection	#Req per Session	IRT's in a Connection	IRT's in a Session
ALL-HTTP	30 M	1.4 M		57890	278K	4	13	2.34	17.23
Google Maps	221 K	1127		204476	2288K	18	197	1.39	1.54
lokalisten.de	128 K	3822	mean	31856	129K	8	34	0.38	4.52
Google Mail	140 K	1020		9742	371K	4	138	23.02	31.84
GMX	288 K	6101		14163	95K	7	47	0.53	4.29
ALL-HTTP				332	688	1	2	0.0987	0.2035
Google Maps				25199	161675	4	21	0.0288	0.0076
lokalisten.de			median	1678	7854	3	7	0.0347	0.0406
Google Mail				3	27932	1	23	4.3735	9.2202
GMX				428	6863	3	29	0.0400	0.0489

orders of magnitude with a median of 332 Bytes and a mean of 58 KB. Some connections are clearly used to transfer a huge number of bytes, e.g, due to downloading some large image or video file embedded within a HTTP page, or a big software package, or when HTTP is used as transport protocol for P2P protocols, such as Bittorrent.

The tails of the AJAX based Selected-4 applications are not as heavy. Yet, except for Google Mail the curves lie on top of the ALL-HTTP traffic for most of the plot which is reflected in the statistics as well, e.g., the median and mean for Google Maps is larger, i.e., 25 KB and 204 KB respectively.

To further explore the differences in the body of the distribution we show the PDF for Google Maps and Mail as well as ALL-HTTP traffic in Figure 4. In general we note that the Selected-4 applications (see for example, Google Maps) transfer more bytes than ALL-HTTP connections. This probably stems from multiple larger image/JavaScript library transfers, when, for example, Google Maps users pan and zoom their map. In particular, only 39.6% of the MWN-07 Google Maps connections comprise of connections that transfer less than 10 KB, whereas 81.8% of the ALL-HTTP connections from MWN-07 transfer less than 10 KB. Similar observations hold for the LBNL-07 data set. Moreover, we note that the shape of the ALL-HTTP connection has not changed substantially over the years if compared with results from 1997 [11].

Google Mail differs and shows a clear spike for 3 bytes requests. This is due to periodic server polling by the client-side AJAX engine of Google Mail. Once we move from HTTP connections to HTTP sessions (Figure 5), this artifact is removed and the probability mass of all Selected-4 applications clearly lies to the right of that for ALL-HTTP traffic. This is reflected in the median but not in all means. But recall that the mean is dominated by the very large transfers within the ALL-HTTP traffic.

We next move to the number of HTTP request within a session. Figures 7 and 8 show the CCDF and PDF for ALL-HTTP and Selected-4 sessions in the MWN-07 data set. These figures highlight the "chatty" nature of the Selected-4 applications - on average they issue many more requests than ALL-HTTP traffic whose first fifty percent of the sessions are limited to 2 requests. Part of these additional requests are due to the Web 2.0 characteristics of the Selected-4 applications while the others are likely due to longer session duration. Interestingly, a look at the PDF reveals that Google Maps issues more requests than the email or social networking applications. A likely explanation is that Google Maps implements pre-fetching more aggressively.

The typical duration of an ALL-HTTP session (Figure 6), is shorter than for AJAX enabled applications. Half of the ALL-HTTP sessions last between 0.008 and 2.13 seconds (5% – 55% quantile across all sessions) while 50% of Google Maps sessions in the MWN-07 data set last between 13.04 seconds and 2 hours and 9 minutes (30% – 80% quantile across all sessions). On the other hand the first period only accounts for 20.7% of the Google Maps session while the second only accounts for 23.87% of the ALL-HTTP traffic. One reason for the longer session duration may be that these specific applications are able to keep the users attention longer than a typical web site. Overall these characteristics indicate that AJAX enabled applications last longer and are more active than ALL-HTTP sessions.

Finally, Figures 9 and 10 show the inter-request-times between requests within a session. The most interesting feature of this density graph is that Google Maps' inter-request-times are very similar and significantly shorter, i.e., more frequent, than for

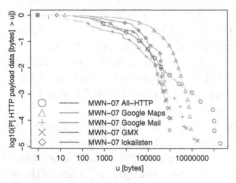

Fig. 3. HTTP payload bytes per connection

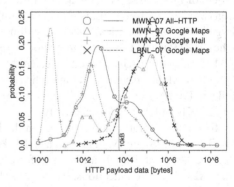

Fig. 4. HTTP payload bytes per connection

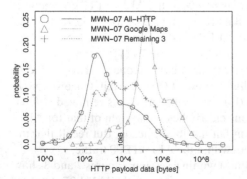

Fig. 5. HTTP payload bytes per Session

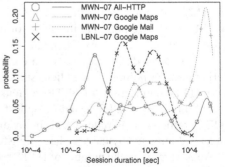

Fig. 6. Session durations

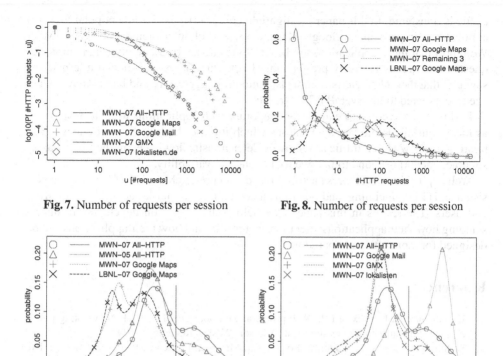

Fig. 7. Number of requests per session

Fig. 8. Number of requests per session

Fig. 9. PDF of inter-request-times within each session: ALL-HTTP and Google Maps

Fig. 10. PDF of inter-request-times within each session: Google Mail, GMX and lokalisten.de

ALL-HTTP for both MWN-07 and LBNL-07. As such the traffic pattern is burstier. Moreover, there has not been a major change for ALL-HTTP from 2005 to 2007. The majority of requests are clearly automatically generated, as they are executed within 1 second (see support line; > 1 second corresponds roughly to human-issued browser request) in all sessions. Google Maps is again the most extreme application. Most likely this is caused by the utilization of pre-fetching for supporting the dynamic features of Google Maps.

Moreover, we note that different service providers can use the AJAX capabilities in different manners. GMX and Google Mail are both Web based email applications. Yet, the inter-request-times differ dramatically. The reason for this is that Google Mail uses a polling interval of roughly 120 seconds (those 3 Bytes requests from Figure 4). Once these are removed the densities are quite similar again.

4   Conclusions

The overall transition of the web from a hyperlinked document repository into a real-time application platform has ramifications for the underlying Internet over which web traffic is transfered. In this paper we highlight characteristics of some popular Web 2.0 applications, in particular - Google Maps, Google Mail, lokalisten.de, and GMX Mail. We report that these applications are heavy (bytes transferred), chatty (many more requests), and greedy (actively pre-fetching data). Our analysis of their traffic patterns suggests that their characteristics translate into more aggressive and bursty network usage as compared to the overall HTTP traffic.

End users have come to expect contemporary web applications to be as responsive as locally installed software applications which imposes high QoS requirements. Yet, treating this new HTTP traffic as relatively deterministic flows (i.e., in the same way as streamed media) is bound to fail due to the inherent variability.

Web application developers have embraced data pre-fetching, HTTP connection persistence, HTTP pipelining, and other advanced features to mask network latency from end users. The results in this paper may help web application developers in understanding how their applications affect Internet traffic, and how their applications can be designed for more efficient operation.

References

1. Krishnamurthy, B., Rexford, J.: Web protocols and practice: HTTP/1.1, Networking protocols, caching, and traffic measurement. Addison-Wesley, Reading (2001)
2. Fielding, R., Gettys, J., Mogul, J., Frystyk, H., Masinter, L., Leach, P., Berners-Lee, T.: Rfc 2616, hypertext transfer protocol – http/1.1 (1999)
3. Zakas, N., McPeak, J., Fawcett, J.: Professional AJAX. Wiley, Chichester (2006)
4. Crovella, P.B.M.: The network effects of prefetching. In: INFOCOM (1998)
5. Abrams, M., Standridge, C.R., Abdulla, G., Williams, S., Fox, E.A.: Caching proxies: limitations and potentials. In: WWW Conference (1995)
6. Barford, P., Bestavros, A., Bradley, A., Crovella, M.E.: Changes in Web client access patterns: Characteristics and caching implications. In: World Wide Web (1999)
7. Challenger, J., Iyengar, A., Danzig, P.: A scalable system for consistently caching dynamic Web data. In: INFOCOM (1999)
8. The impact of AJAX on web operations (2005), http://www.bitcurrent.com/?p=105
9. Document Object Model (DOM) (2007), http://www.w3.org/DOM
10. Paxson, V.: Bro intrusion detection system (2007), http://www.bro-ids.org
11. Feldmann, A., Rexford, J., Caceres, R.: Efficient policies for carrying Web traffic over flow-switched networks. IEEE/ACM Trans. Networking 6(6) (1998)

Measurement and Estimation of Network QoS Among Peer Xbox 360 Game Players

Youngki Lee[1], Sharad Agarwal[2], Chris Butcher[3], and Jitu Padhye[2]

[1] KAIST
[2] Microsoft Research
[3] Bungie Studios

1 Introduction

The research community has proposed several techniques for estimating the quality of network paths in terms of delay and capacity. However, few techniques have been studied in the context of large deployed applications. Network gaming is an application that is extremely sensitive to network path quality [1,2,3]. Yet, the quality of network paths among players of large, wide-area games and techniques for estimating it have not received much attention from the research community.

Network games broadly fall into two categories. In some games (e.g. MMORPGs, web-based casual games, Quake) with a client-server architecture, players communicate with a large, *well-provisioned*, and *dedicated* game server [4,5]. In some games with a peer-to-peer (P2P) architecture, players communicate with each other directly or via a dynamically chosen peer at some player's house. In Ghost Recon, Halo series, and others for the Xbox and Xbox 360 consoles, a server assists players in discovering other peers to host and play with.

Accurate and scalable estimation of the network path quality (NPQ) between peer game players is especially critical for games with a P2P architecture. These players need to have good network connectivity to each other, so accurate NPQ data is essential for "matchmaking" - i.e. to determine which players should play with each other. Furthermore, NPQ estimation needs to be done in a scalable manner. If the number of peers is large, it may not be not feasible to probe all of them.

Prior research on P2P games has used data from only a small number of players [6]. We study a much larger data set, from Halo 3 : a popular Xbox 360 game. We cover 5.6 million unique IP addresses that played 39.8 million game sessions in 50 days. Peers in each game session gather NPQ data and report it to the central Xbox server for matchmaking purposes.

This paper makes the following contributions:

- We present data from a large P2P gaming application. The population is several orders of magnitude larger, and far more geographically diverse than any previously reported study. Given the number and geographical diversity of players, we consider this to also be a large-scale study of path quality over the wide-area Internet.
- We study temporal and geographical correlations in the NPQ data, and propose three different predictors that can provide a rough estimate of NPQ between a pair of players, without requiring any probing. There can be millions of game players

M. Claypool and S. Uhlig (Eds.): PAM 2008, LNCS 4979, pp. 41–50, 2008.
© Springer-Verlag Berlin Heidelberg 2008

on-line at any time, and any techniques that can avoid having to perform network probes between all of them can not only reduce network overhead but also reduce the amount of time players have to wait before starting a game over the Internet.

2 Background

The Microsoft Xbox 360 game console supports on-line game play with the Xbox Live subscription service. The Halo series of First Person Shooter (FPS) games has sold over 15 million copies worldwide. We focus on the latest edition, Halo 3. Each Halo 3 Internet game session can include up to 16 players. One console in each game session is selected to be the game host or server. All game communication between other players is relayed through this host console. The Xbox Live server provides accounting and matchmaking services. Therefore, the NPQ between consoles and the Xbox Live server is less important to the overall gaming experience than the NPQ between the consoles themselves. An "excellent" Halo 3 experience has under 50ms of latency and 50Kbps to 70Kbps of bandwidth between each client console and the host console. Note that the host console may consume up to 1Mbps ((16-1)*70Kbps) of bandwidth. A "good" experience can be achieved with 150ms latency and 30Kbps of bandwidth. Hence, it is important to group consoles so that they each have good NPQ to the host console. This is critical in this architecture because the host is a fellow player's console, typically on a consumer broadband connection, and not a well provisioned, dedicated server.

The Xbox Live server helps with "matchmaking" - setting up such groups, of up to 16 players, from among the hundreds of thousands on-line at any time. A player who starts an Internet game session will sign on to the Xbox Live service and run Halo 3 on her console. She will select Internet game play and can specify several criteria for the session, such as the type of game (e.g. free for all or team objective). With some probability, this console will become a peer game host, instead of a game client. This probability depends on the chosen type of game. If the console is a game host, it will wait for other consoles to discover it, probe their NPQ to it, and join the game.

If it is a game client, Xbox Live will send it IP addresses for the other consoles on the Internet that are hosting games of the specified type. This console will send and receive several packet pairs with each IP address. The Xbox 360 networking stack implements the standard packet pair estimation technique [7]. Packet pairs are performed serially and do not overlap with each other. The console will then have an estimate of the round-trip latency (RTT), and the upstream and downstream network capacity with each candidate game host. While being very lightweight, packet pair measures bottleneck link capacity but not available bandwidth. These values are logged by the Xbox Live service. The user is shown those host consoles that it has the best NPQ to. For conciseness, we leave out several details such as NAT traversal.

Little is known about the population of on-line P2P game players. Their geographic diversity, diurnal behavior, typical network delay and capacity are useful parameters to network models of game systems for future research. This information can help build estimators of NPQ between any two game consoles on the Internet. Even merely identifying the pairs of consoles with extremely poor NPQ can significantly reduce the total number of probes, thereby reducing network overhead and user wait time.

3   Data and Methodology

Xbox Live stores information about every Internet game session for Halo 3. In a typical week ending on 29 January 2008, we find that 72.5% of Internet game sessions required matchmaking; when weighted to account for players per game, it is 83.5%. By a "session", we mean an attempt to search for an Internet game - the user may have eventually joined a game or decided not to. The log has the UTC time and the public IP address of the console searching for a game. This console may have probed several other consoles that were hosting games of the requested type - for each probe to a candidate host console, we have the host IP address, median round trip time (RTT), and average capacities upstream to host and downstream from host. We use the term "probe" to mean 4 packet pair tests from the client console to a host console and 4 in the reverse direction. We use "player", "console" and "IP address" interchangeably.

Table 1. Data sets

Halo 3 Phase	Start	End	Distinct IPs	Matchmaking games	Hosts probed
Internal alpha	11/30/2006	01/23/2007	4,025	314,606	207,595
Internal beta	05/08/2007	05/21/2007	732,487	20,747,695	33,338,060
Public beta	05/22/2007	06/11/2007	903,782	23,182,323	38,453,035
Release	11/14/2007	01/03/2008	5,658,951	39,803,350	126,085,887

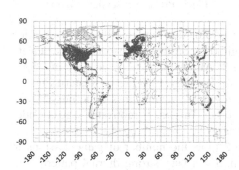

Fig. 1. Geographic distribution of players

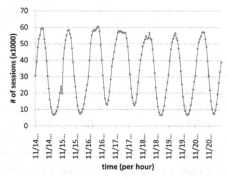

Fig. 2. Game sessions per hour

Table 1 lists our data sets. For conciseness, we focus on the "Release" data set for Halo 3. Due to the extremely large number of game plays, we limit the data set in two ways - we consider a 50 day period and we only consider a randomly selected 20% of the matchmaking game sessions. The resulting data set covers over 126 million probes between over 5.6 million IP addresses. For geographic analysis, we use the commercial MaxMind GeoIP City Database from June 2007. It was able to provide the latitude and longitude for over 98% of these IP addresses.

4   Player Population Characterization

In this section, we analyze the basic characteristics of the player population, such as the geographic distribution of the players, when and how often they play the game. We also look at the overall NPQ data such as distributions of RTT and capacity.

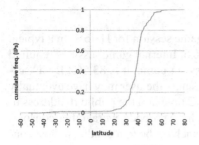
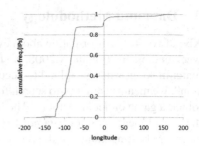

Fig. 3. Latitude and longitude density of players

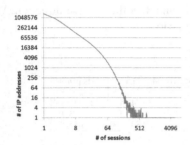
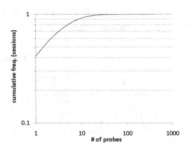

Fig. 4. Game sessions per IP address (log-log)   **Fig. 5.** Probes per session (log-log)

Figure 1 shows the geographic locations of all 5,658,951 unique IP addresses, which correspond to 68,834 unique latitude and longitude coordinates. To examine the density of players in each region, we present Figure 3. Almost 85% of players are in USA - longitudes -130 to -60, and latitudes 30 to 50. Roughly 15% are in western Europe. Since players are spread across this large geographic region, it is quite possible for consoles that are "too far apart" to probe each other. This partly motivates us to consider estimation techniques that will identify such cases before probing.

To see when games were played, Figure 2 plots the number of game sessions in each hour over a representative week. We notice a very strong diurnal pattern with peaks in the evenings in North American time zones - this is not unexpected given the high density of players in USA. We examine game playing behavior in more detail in Figure 4. The number of games attempted from each IP address almost follows a Zipf distribution. In the far right of the graph, one IP address attempted 5438 sessions - over a 50 day period, this is a huge number of games for any 1 individual! We suspect that the IP addresses in this area of the graph are for proxies with many players behind them.

Figure 5 shows a CDF of the number of consoles hosting a game that were probed in each session. While there are many sessions that probed few consoles, there are some that probed as many as 400 consoles. This number depends on how many game hosts the Xbox Live server gives a console requesting a game, which in turn depends on how many consoles are available at the time and the type of game requested.

Now we consider overall NPQ data. Figure 6 shows the CDF of RTT across all probes. Over 25% of the measurements are above 150ms, which is an upper bound for a responsive user experience in typical FPS games [1]. We want to pre-determine in which cases the RTT will be above 150ms and skip probing altogether, thereby

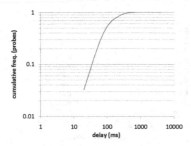
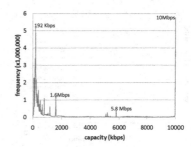

Fig. 6. RTT delay reported by probes (log-log)   **Fig. 7.** Downstream capacity reported by probes

potentially reducing the total number of probes by 25%. Figure 7 shows the distribution of measured capacity across all probes, in the direction from the console hosting a game to the console requesting to join it. The graph for upstream capacity is similar. We see peaks around typical capacities for broadband access in USA (e.g. 192Kbps, 1.5Mbps), within some marginal error due to the packet pair estimation technique.

5   NPQ Prediction

The NPQ probing technique that Halo 3 uses consists of 16 packets per console being probed (4 packet pairs in each direction). However, there can be many candidate hosts for a game. For scalability and to minimize user wait time, we want to reduce the total number of probes and hence propose the use of NPQ predictors. Our goal is to estimate apriori if a console has good NPQ to a remote candidate host console, without doing a probe. If bad NPQ is predicted, then this candidate should not be probed. If good NPQ is predicted, then limited probing can be done (e.g. only 1 packet pair). If no prediction can be made, standard probing should ensue. Based on our analysis of the NPQ data, we now propose and evaluate three NPQ predictors.

5.1   IP History Predictor

We hypothesize that a probe between a pair of consoles at time t_1 produces an NPQ estimate that is still valid at a later time t_2. This may be true if the median RTT and average upstream and downstream bottleneck capacities do not vary significantly over a period of $\delta = t_2 - t_1$. To test this hypothesis, and estimate how large δ can be, we examine NPQ data for pairs of IP addresses over different periods of time.

Figure 8 shows the CDF of the coefficient of variation (CV) in RTT for pairs of IP addresses over different time windows. For instance, the "Within 5 min" line shows the CV of RTTs from probes between the same pair of IP addresses within any 5 minute window. To draw meaningful conclusions, we consider only those IP pairs that probed each other at least 5 times during that period. We have plotted similar lines for 30 minutes, 6 hours, 1 day and the entire 50 day trace (the line labeled "no constraints"). The lines for all 5 time windows overlap with each other. For over 90% of IP address pairs that probed each other at least 5 times, the variation in RTT estimates was minuscule (CV under 0.2), even up to a δ of 50 days. For comparison we plot the "baseline" - instead of considering CV for each pair of IPs, we consider the CV for each single IP.

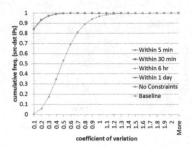

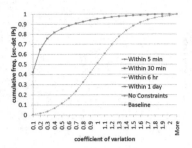

Fig. 8. RTT variation for IP pairs   **Fig. 9.** Capacity variation for IP pairs

That is, for each IP address, the CV across all RTT estimates where this IP address was involved, across the entire trace. This line is well below the others, indicating that the RTTs are spread across a wide range. We conclude that the IP History Predictor can perform quite well for predicting RTT, and the δ can be as large as 50 days.

Figure 9 has the same graph for downstream capacity. The upstream capacity graph is very similar and is omitted for conciseness. Again, δ does not affect the NPQ prediction. While the CV is larger, it is under 0.65 for 90% of the IP pairs, and is still much higher than the "baseline". Thus we believe the IP History Predictor adequately predicts the NPQ between a pair of consoles based on an NPQ estimate from a prior probe.

5.2   Prefix History Predictor

We have shown the IP History Predictor to work only when pairs of consoles have probed each other in the past. This may reduce the number of probes in only a limited set of cases. Thus we now propose the Prefix History Predictor - this is similar to the IP History predictor, except it uses IP *prefix* pairs. We hypothesize that a probe between a pair of consoles A_1 and B_1 at time t_1 produces an NPQ estimate that is still valid at a later time t_2 for a different pair of consoles A_2 and B_2, as long as A_1 and A_2 belong to one BGP prefix, and B_1 and B_2 belong to one BGP prefix.

This predictor may be accurate if consoles in the same prefix share similar last mile access. However, broadband ISPs offer several access packages (e.g., 192Kbps DSL or 1.5Mbps DSL), and the prefix may indicate geographic location more than link speed. Thus, predictions for capacity may be less accurate than for RTT. We now analyze NPQ data for pairs of IP prefixes that probed each other at least 5 times. We find a console's prefix by a longest prefix match against the 12/27/2007 RouteViews BGP table [8].

Figure 10 shows the performance of this predictor for delay, and can be compared to Figure 8. When considering prefix pairs, δ has a bigger impact - the older the original probe, the worse the prediction. Since CV is a relative metric, small variations in small RTTs (e.g. 5ms versus 10ms) can produce a large CV. Thus in Figure 11 we look at the semi-interquartile range (SIQR) of RTT estimates for prefix pairs for no limit on δ (i.e., the "no constraints" case). For 90% of prefix pairs, the SIQR is under 40ms. Thus it is the outliers beyond the 25%-75% SIQR that contribute to this additional variability.

Figure 12 shows the performance of this predictor for downstream capacity estimation. For δ beyond 5 minutes, it is a very poor predictor. We suspect this is due to different subscription levels of last mile capacity within the same prefix.

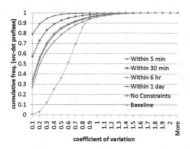

Fig. 10. RTT variation for prefix pairs

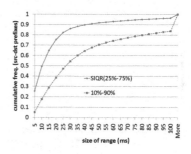

Fig. 11. SIQR of RTT for prefix pairs

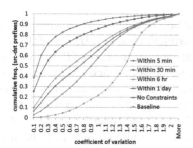

Fig. 12. Capacity variation for prefix pairs

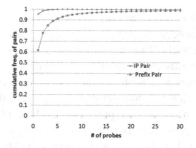

Fig. 13. Number of repetitive probes

Based on these results, we believe the Prefix History Predictor can adequately predict the RTT between a pair of consoles based on an NPQ estimate from a prior probe between consoles in the same pair of prefixes. However, this prediction is not as accurate as the IP History Predictor - so we suggest first applying that predictor, and only if it cannot make a prediction, then using the Prefix History Predictor. To show in how many cases this would apply, we present Figure 13. We plot the CDF of the number of repeated probes in the entire trace between the same pair of IP addresses, and the same pair of prefixes. Only about 5% of pairs of consoles probed each other more than once, while about 39% of prefix pairs probed each other more than once. Note that we have clipped the horizontal axis at 30 for presentation purposes - the IP pair line continues to 114 probes, and the prefix pair line continues to 14,513.

5.3 Geography Predictor

While 39% of prefix pairs is still a significant fraction of the number of consoles, and has the potential to reduce a far larger portion of probes those prefixes probed each other several times, there is still about 61% of prefix pairs left. We now consider the Geography Predictor - we hypothesize that the geographic distance between two consoles has a strong correlation with their RTT, and that current databases for mapping IP addresses to geographic coordinates are reasonably accurate for this. This may be true if distant IP addresses traverse longer links and more router hops to communicate. This predictor does not consider past history, and thus could be applied to any pair of consoles.

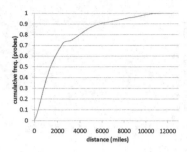
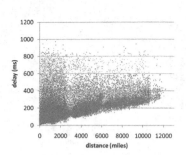

Fig. 14. Probe distance distribution   **Fig. 15.** Distance-RTT correlation

It is not clear why geographic distance would be correlated with down/up stream bottleneck capacity - our analysis indicates the correlations are -0.075 and -0.097, respectively. Thus we omit detailed results on capacity for conciseness and focus on RTT.

We use a MaxMind location database to convert the source and destination IP addresses in a probe to the geographic coordinates, and apply the great circle distance algorithm [9] to calculate distance. The distribution in Figure 14 shows a wide range of distances between probe hosts. About 14% of probes traversed over 5,000 miles, which indicates there is room for optimization by filtering out these console pairs from the probe list. The graph also shows that we have enough samples to examine the correlation between distance and delay.

Figure 15 plots the correlation between distance and RTT for 100,000 randomly selected probes (we were unable to plot over 126 million points on the same graph – different samples of 100,000 points gave us very similar graphs). We see a very strong correlation between the geographic distance and minimum RTT. However, there is a lot of noise above that minimum, which may be due to queuing delays and sub-optimal routes. We conclude that the Geography Predictor is useful for filtering out pairs of IP addresses that are too far apart to have a low RTT.

5.4   Using Predictors in Matchmaking

Incorporating these three predictors into matchmaking is not difficult. For the IP History Predictor, each console will keep a history of previous probes that it was involved in. It can look up this history before attempting any future probes, and decide which candidate game hosts to ignore. For the Prefix History Predictor, the Xbox Live server can filter the set of candidate game hosts it provides to each console based on their prefixes and past probe history. The server already has the past NPQ estimates, and it can easily keep fresh BGP tables to look up prefixes. The Geography Predictor requires an IP to geographic coordinate database, on either the Xbox Live server or on each console.

6   Prior Work

Most prior work on network gaming has focused on games with a client-server architecture [4,5,10] where the server is well-provisioned and dedicated. The literature on P2P games is very limited. In [6], the authors examine game clients deployed in three access networks: dial-up, cable and ADSL. However, their experiments are limited to one client per access network, and use only one cable and one ADSL link. The game traffic of Halo 2 is analyzed in [11] in a LAN environment for traffic modeling and not for end-to-end network characteristics between real Halo 2 players.

There has been much prior work on efficient and accurate NPQ prediction. For conciseness, we identify those done in the context of network gaming. As before, most of this work is for client-server games. In [12], a simple server-side method is proposed to improve server location discovery and reduce probe traffic. Our NPQ prediction methods focus also on reducing overall probe time since that directly affects user wait time. Also, we not only utilize the geographic location of consoles but also previous probe results. A flooding-style server discovery mechanism is proposed in [13] to quickly locate local servers and prevent single directory server failure. That does not scale to P2P games, since in our case several hundreds of thousands of consoles can be on-line at any time. A server selection algorithm is proposed in [14] for distant game players who want to play specifically with each other. Our work considers the general case of joining players to any acceptable game, and thus considers NPQ data and correlators across all on-line consoles. The geographic distribution of game servers and players is used in [15] to redirect players to close game servers. While [16] does not consider on-line games, they correlate geographic location and network delay to find a host, and their experimental result about the correlation complements ours.

Outside network games, there has been a lot of research on characterizing NPQ over the Internet. Many of these [17,18] use PlanetLab nodes. They are mostly located in high-performance and stable academic networks, and thus do not reflect the characteristics of consumer access networks. In [19], the constancy of NPQ over time among 31 hosts is studied within a stable academic network. Our work significantly complements prior work in terms of scale and diversity of network connectivity. Studies of hosts behind consumer broadband lines are rare. It is extremely difficult to build a large testbed of such hosts on the Internet. While [20] characterizes network performance between consumer broadband hosts, they use only 25 hosts. More recently, [21] studies residential network link capacities, RTT, and loss rates through relatively large-scale measurement studies. They use 1,894 hosts behind 11 major DSL and cable providers in North America and Europe. Our study is much larger in scale, involving over 5.6 million hosts. Furthermore, they do not characterize direct network connections between pairs of broadband hosts since they measure from several vantage points located in academic networks. Techniques for estimating NPQ have been studied extensively [7,22]. Our work focuses not on the techniques itself, but on the NPQ data.

7 Conclusions

We studied the quality of network paths among Xbox 360 game consoles playing Halo 3. We focused on network delay and capacity measured between players prior to each Internet game match-up. We studied the general characteristics of the player population such as geographical diversity and diurnal patterns of game play. We leveraged our understanding of these characteristics to propose three predictors for determining path quality without additional probe traffic : IP and prefix history-based and geography-based. Our evaluation of these predictors showed that they can significantly reduce the number of probes and hence user wait time during matchmaking. For future work, we plan on comparing the initial NPQ estimate to actual in-game network performance.

References

1. Dick, M., Wellnitz, O., Wolf, L.: Analysis of factors affecting players performance and perception in multiplayer games. NetGames (2005)
2. Quax, P., Monsieurs, P., Lamotte, W., Vleeschauwer, D.D., Degrande, N.: Objective and subjective evaluation of the influence of small amounts of delay and jitter on a recent first person shooter game. NetGames (2004)
3. Armitage, G.: An experimental estimation of latency sensitivity in multiplayer Quake 3. In: ICON (2003)
4. Feng, W., Chang, R., Feng, W., Walpole, J.: Provisioning on-line games: A traffic analysis of a busy Counter Strike server. In: IMW (2002)
5. Kim, J., Choi, J., Chang, D., Kwon, T., Choi, Y., Yuk, E.: Traffic characteristics of a massively multi-player online role playing game. NetGames (2005)
6. Jehaes, T., Vleeschauwer, D.D., Coppens, T., Doorselaer, B.V., Deckers, W.N.E., Spruyt, J., Smets, R.: Access network delay in networked games. NetGames (2003)
7. Carter, R.L., Crovella, M.E.: Measuring bottleneck link speed in packet-switched networks. Technical report, Boston University (March 1996)
8. University of Oregon: Routeviews project page http://www.routeviews.org/
9. Hexa software development center: Distance calculation method between two latitude and longitude coordinates, http://zipcodeworld.com/docs/distance.pdf
10. Chambers, C., Feng, W., Sahu, S., Saha, D.: Measurement-based characterization of a collection of on-line games. In: IMC (2005)
11. Zander, S., Armitage, G.: A traffic model for the Xbox game Halo2. In: NOSSDAV (2005)
12. Zander, S., Kennedy, D., Armitage, G.: Server-discovery traffic patterns generated by multiplayer first person shooter games. NetGames (2005)
13. Henderson, T.: Observations on game server discovery mechanisms. NetGames (2003)
14. Gargolinski, S., Pierre, S., Claypool, M.: Game server selection for multiple players. NetGames (2005)
15. Chamber, C., Feng, W., Feng, W., Saha, D.: A geographic redirection service for on-line games. ACM Multimedia (2003)
16. Fdida, S., Duarte, O.C.M.B., de Rezende, J.F., Ziviani, A.: Toward a Measurement-Based Geographic Location Service. In: Barakat, C., Pratt, I. (eds.) PAM 2004. LNCS, vol. 3015, pp. 43–52. Springer, Heidelberg (2004)
17. Lee, S.-J., Basu, S., Sharma, P., Banerjee, S., Fonseca, R.: Measuring Bandwidth Between PlanetLab Nodes. In: Dovrolis, C. (ed.) PAM 2005. LNCS, vol. 3431, pp. 292–305. Springer, Heidelberg (2005)
18. Banerjee, S., Griffin, T.G., Pias, M.: The Interdomain Connectivity of PlanetLab Nodes. In: Barakat, C., Pratt, I. (eds.) PAM 2004. LNCS, vol. 3015, pp. 73–82. Springer, Heidelberg (2004)
19. Zhang, Y., Duffield, N., Paxson, V., Shenker, S.: On the constancy of Internet path properties. In: IMW (2001)
20. Lakshminarayanan, J., Padmanabhan, V.N.: Some findings on the network performance of broadband hosts. In: IMC (2003)
21. Dischinger, M., Haeberlen, A., Gummadi, K.P., Saroiu, S.: Characterizing residential broadband networks. In: IMC (2007)
22. Dovrolis, C., Ramanathan, P., Moore, D.: Packet-dispersion techniques and a capacity-estimation methodology. IEEE/ACM Transactions on Networking (December 2004)

Evaluation of VoIP Quality over WiBro

Mongnam Han[1], Youngseok Lee[2], Sue Moon[1], Keon Jang[1], and Dooyoung Lee[1]

[1] Computer Science Department, KAIST
[2] School of Computer Science & Engineering, Chungnam National University

Abstract. In this work, we have conducted experiments to evaluate QoS of VoIP applications over the WiBro network. In order to capture the baseline performance of the WiBro network we measure and analyze the characteristics of delay and throughput under stationary and mobile scenarios. Then we evaluate QoS of VoIP applications using the E–Model of ITU–T G.107. Our measurements show that the achievable maximum throughputs are 5.3 Mbps in downlink and 2 Mbps in uplink. VoIP quality is better than or at least as good as toll quality despite user mobility exceeding the protected limit of WiBro mobility support. Using RAS and sector identification information, we show that the handoff is correlated with throughput and quality degradation.

1   Introduction

Recent emerging wireless networks such as 3G cellular and wireless LAN (WLAN) allow users choices in accessing the Internet based on one's need and cost. WLAN with a high data rate (up to 54Mbps) supports low mobility and limited coverage. Cellular networks support high mobility with low bandwidth. The broadband wireless access (BWA) systems address the market between WLAN and cellular networks. Their goal is to support higher bandwidth than 3G cellular networks, but less mobility for mobile end-user devices. The IEEE 802.16 family of standards specifies the air interface of fixed and mobile BWA systems. WiMax is a subset of the 802.16 standards whose main goal is product compatibility and interoperability of BWA products, just as WiFi is to the 802.11 standards. WiBro has been developed as a mobile BWA solution in Korea, and is generally considered a precursor to WiMax. It is a subset of consolidated version of IEEE Standard 802.16-2004 (fixed wireless specifications), P802.16e (enhancements to support mobility), and P802.16-2004/Cor1 (corrections to IEEE Standard 802.16-2004). The profiles and test specifications of WiBro will be harmonized with WiMAX Forum's mobile WiMAX profiles and test specification, drawing a convergence of the two standards.

Today's Internet users not only write emails and surf the web, but also make Voice over IP (VoIP) calls, play online games, and watch streaming media. These real-time applications have stringent Quality of Service (QoS) requirements on delay and loss. WiMax and WiBro standards have defined multiple service types in order to guarantee different levels of QoS. However, at the initial phase of deployment, often only the best-effort service is made available, while users do

M. Claypool and S. Uhlig (Eds.): PAM 2008, LNCS 4979, pp. 51–60, 2008.

not limit themselves to emails and web surfing over emerging wireless technology networks.

In this work, we conduct experiments to evaluate QoS of VoIP applications over the WiBro network. In order to capture the baseline performance of the WiBro network, we measure and analyze delay, loss, and throughput of constant bit rate streams in both stationary and mobile scenarios. We have measured maximum throuhgputs of 5.3 Mbps on downlink and 2 Mbps on uplink. Packet loss and throughput exhibit more variability in the mobile scenario than stationary. Then we evaluate QoS of VoIP applications using the E–Model of ITU–T G.107 also in both stationary and mobile scenarios. VoIP quality is better than or at least as good as toll quality even in the mobile scenario. By combining the packet traces with physical layer information, we show that the handoff is correlated with throughput and quality degradation on VoIP quality. We note that the deployed WiBro network is lightly loaded.

The rest of this paper is organized as follows. In section 2, we describe the background and related work. In Section 3, we describe our measurement experiment setup in a WiBro network and present the VoIP quality evaluation methodology. We present our analysis results in Section 4 and wrap up the paper with a summary in Section 5.

2   Background and Related Work

Fixed WiMax was first used to assist in the relief effort for the 2004 tsunami in Aceh, Indonesia, and now has more than 350 service providers around the world [14]. WiBro, a mobile BWA service, had its public demonstration in December 2005, and has been in service since June 2006 in Korea. The network architecture of WiBro in the phase I standardization [13] is shown in Figure 1. The WiBro network consists of Access Control Routers (ACR), Radio Access Stations (RAS), Personal Subscriber Stations (PSS), and the network service provider's IP network. An RAS is the interface between PSSs and the core network at the physical layer and it also controls the radio resource at the data link layer in conjunction with an ACR. One of the distinguishing features of WiBro from cellular networks is that Internet Protocol (IP) is used beween an ACR and RASs and also between ACRs. WiBro uses Time or Frequency Division Duplexing (TDD or FDD) for duplexing and Orthogonal Frequency Division Multiple Access (OFDAS) for robustness against fast fading and narrow-band co-channel interference. So far, five service types have been proposed and incorporated into 802.16e: unsolicited grant service (UGS), real-time polling service (rtPS), extended real-time polling service (ertPS), non-realtime polling service (nrtPS), and best effort service (BE). However, only BE is used in current deployment in Korea.

Ghosh et al. use a link-level simulation to analyze the 802.16 fixed WiMax system [4]. Cicconetti et al. analyze the effectiveness of the 802.16 MAC protocol for supporting QoS by simulation and evaluate various scheduling algorithms [1]. In [12], the authors evaluate UGS, rtPS, and ertPS scheduling algorithm in IEEE

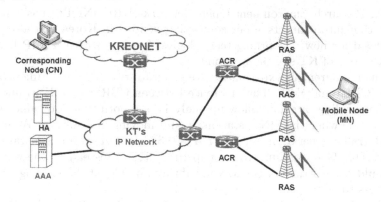

Fig. 1. Experimental environment over WiBro

802.16e system in OPNET simulation. Most of prior work focuses on investigating the performance at the physical layer and MAC layer largely through either simulation or experiment with limited mobility. In this work, we focus on the end-to-end performance at the application layer considering mobility in real life. Our work is unique in that we focus on empirical measurements from a real deployed network, world's first commercial deployment of the mobile WiMax technology.

3   Experiment Setup and Evaluation Methodology

We begin this section with a description of our measurement experiment setup. Then in Section 3.2, we describe the ITU-T E-model for VoIP quality evaluation.

3.1   WiBro Performance Measurement

In Korea, KT launched WiBro coverage for nine subway lines in Seoul on April, 2007. The Seoul subway system moves millions of people a day through an extensive network that reaches almost all corners within the city and major satellite cities outside. The maximum speed of Seoul subway trains is 90 km/h, and it takes about 1~2 minutes between two stations. We have considered measurement experiments in vehicles moving at or under 60 km/h, the upper limit of WiBro, but chosen the subway, as it presents a more popular scenario with users. Commuters in subway are more likely to use mobile devices than those in moving vehicles, as the measurement experiment on a subway train is easier for us. We have conducted our measurement experiments on subway line number 6. It has 38 stations over a total distance of 35.1 km and six RASs.

We have placed a mobile node (a laptop with a WiBro modem) in the WiBro network and installed a stationary node (a desktop PC) connected to the Internet over a fixed line so that we could focus on the WiBro network performance. We refer to the laptop as the Mobile Node (MN) and the PC as the Corresponding Node (CN), and mark them as such in Figure 1. In order to place the CN as close to the WiBro network as possible, we use a PC directly connected to a router on Korea Research Environment Open Network (KREONET). It is a research network that interconnects super computing centers in Korea and also is used as a testbed for new networking technologies. It peers with KT's IP backbone network at one of KT's exchange points.

For our measurement experiments, we generate two types of traffic: constant bit rate (CBR) and VoIP. The difference between CBR and VoIP traffic lies in the packet sending rate and follow-up analysis. For both types of traffic, we take measurements when the MN is stationary and moving in a subway. We use *iperf* for CBR traffic generation [6], and *D-ITG* [3] for VoIP traffic generation. We configure D-ITG to measure round-trip time (RTT) instead of one-way delay, as we could not instrument the MN in subway and CN at an exchange point to have access to GPS-quality clock synchronization.

Multiple types of handoff are possible in the WiBro network. An inter-ACR handoff takes longer than inter-RAS or inter-sector handoff. An inter-sector handoff is between two sectors within an RAS. An RAS typically has three sectors. Using a custom tool developed to monitor inter-sector and inter-RAS handoffs, we collect RAS identifiers and corresponding sector identifiers. By aligning the changes in RAS and sector identifiers with the measurement data, we can pinpoint the moments of handoffs in our data.

3.2   VoIP Quality Evaluation

The classic way to evaluate speech quality is Mean Opinion Score (MOS) [9]. However, it is time consuming, costly, and not repeatable, as human experts are involved in the evaluation. Perceptual Speech Quality Measure (PSQM) [10] and Perceptual Evaluation of Speech Quality (PESQ) [11] are the most common objective measurement methods for voice quality. Both still require a reference signal to compare a degraded speech signal against and predict a MOS value. They are called psychoacoustic models. The ITU–T E-model does not depend on a reference signal, but uses a computational model to predict voice quality directly from network measurements. The output of the E-model is a single value, called an "R–factor", derived from delays and equipment impairment factors. The ITU–T G.107 [7] defines the relationship between the R factor and MOS as below:

$$MOS = \begin{cases} 1, & \text{For } R \le 0 \\ 1 + 0.035R + R(R - 60) \cdot (100 - R) \cdot 7 \cdot 10^{-6} & \text{For } 0 < R < 100 \\ 4.5, & \text{For } R \ge 100 \end{cases} \quad (1)$$

The E-model is based on a mathematical algorithm. Its individual transmission parameters are transformed into different individual "impairment factors" that are assumed to be additive on a psychological scale. The algorithm of the E-model also takes into account the combination effects for those impairments in the connection which occur simultaneously, as well as some masking effects.

The R-factor calculated by the E-model ranges from 0 (poor) to 100 (excellent) and can be obtained by the following expression,

$$R = R_o - I_s - I_d - I_{e-eff} + A, \quad (2)$$

where

R_o :   Basic signal-to-noise ratio

I_s :   All impairments that occur more or less
simultaneously with the voice signal

I_d :   Delay impairment factor

I_{e-eff} : Effective equipment impairment factor
caused by low bit-rate codec and
by packet loss on the network path

A :   Advantage factor

Cole et al. has reduced (2) to (3) after taking default values for those parameters other than delay and loss [2].

$$R = 94.2 - I_d - I_{e-eff} \tag{3}$$

In this paper, we use (3) in our WiBro VoIP quality and apply Equations (5) and (10) of [2] to translate one-way delay d and loss rate e to I_d and I_{e-eff}.

$$I_d = 0.024d + 0.11(d - 177.3)H(d - 177.3) \tag{4}$$

$$\text{where } H(x) = 0, \text{if } x < 0, \text{ and } H(x) = 1, \text{if } x \geq 0 \tag{5}$$

$$I_{e-eff} = 0 + 30ln(1 + 15e) \tag{6}$$

4   Analysis

On October 5th and 6th, 2007, we took CBR and VoIP measurements in Seoul. For stationary experiments, we placed the MN on KAIST Seoul campus. For mobile experiments, we rode the Seoul subway line 6. For traffic logging, we used *windump* at both the MN and CN. For the VoIP experiments, the MN and CN also dumped log files including sequence numbers, packet departure times, acknowledgement arrival times, and calculated round trip time. The complete set of CBR and VoIP experiments are listed in Table 1

Table 1. The summary of CBR and VoIP experiments (upload/download)

Type	Environment	No of Exps.	Duration (sec)	Rate (Kbps)
CBR	Stationary	55 / 55	120	1500~2500 / 5000~6000
	Stationary	10 / 10	300	2000 / 5300
	Mobile	10 / 10	300	2000 / 5300
VoIP	Stationary	10 / 10	300	64 / 64
	Mobile	10 / 10	300	64 / 64

4.1   CBR Traffic Analysis

In order to capture the baseline performance of the WiBro network, we first measure the maximum achievable throughput. We generated 5 Mbps up to 6 Mbps and 1.5 Mbps up to 2.5 Mbps traffic in quantums of 100 Kbps for download and upload, respectively, and found the bandwidth capped at about 5.3 Mbps downlink and 2 Mbps uplink.

Then we set the transmission rate of our CBR traffic at 5.3 Mbps for downlink and 2 Mbps for uplink with the packet size of 1460 bytes and saturated the link. We conducted 10 sets of 300-second-long uploads and downloads. Due to limited space, we present only the downlink performance. We first plot the throughput of CBR traffic over time and plot it in Figure 2. From 10 300-second-long sets, we get a time span of 3000 seconds, which is the range on the x-axis. We use a 5-second interval to compute the throughput. In Figure 2(a), we see that the throughput remains almost constant when the MN is stationary. When the MN is mobile, the throughput fluctuates. We plot the inter-quartile dispersion of throughput of both the stationary and mobile experiments in Figure 2(b). In the stationary experiment, the inter-quartile range is so small that most 5-second throughput values converge to 5.3 Mbps. In the mobile experiment, the inter-quartile range spans from 4.1 Mbps to 5.1 Mbps, and has noticeably more points below 3 Mbps. To view the dispersion of throughput in a more visually intuitive way, we plot the variability in Figure 2(c) using the second-order difference plot. The difference between two consecutive throughputs are plotted against that between next two consecutive values. In this figure, the median from the center of the mobile station is about 13 times larger than that of the stationary station. The throughput of MN still remains consistently above 1 Mbps.

Next, we analyze the jitter and loss rates of CBR traffic. For this work, we define jitter as the difference between sending intervals and arrival intervals. Figure 3(a) depicts a cumulative distribution function (CDF) of CBR traffic jitter. In both stationary and mobile experiments, more than 90% of jitters are below 15 milliseconds. Given that our traffic by itself saturated the link, this result is encouraging for real-time applications. Now we look at the loss rate of our CBR traffic. In Figure 3(b) the loss rate in the mobile environment is much higher than in stationary. In a WiBro network, a MAC layer retransmission mechanism called Hybrid Auto Repeat reQuest (HARQ) is adopted to reduce loss rate at the cost of increased delay. As our CBR traffic used UDP as an underlying transport protocol and saturated the link, we expect the loss rates to decrease once we lower the sending rate. We revisit the discussion of the loss rate in the next section.

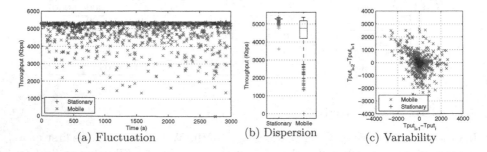

(a) Fluctuation   (b) Dispersion   (c) Variability

Fig. 2. Analysis of CBR traffic throughput over WiBro

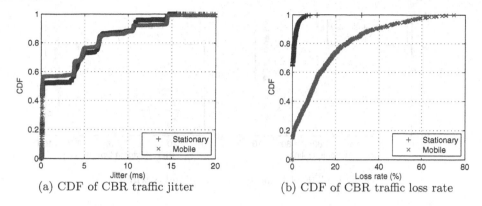

(a) CDF of CBR traffic jitter   (b) CDF of CBR traffic loss rate

Fig. 3. Analysis of CBR jitter and loss over WiBro

4.2 VoIP Traffic Analysis

For VoIP experiments, we have generated voice traffic that has the same characteristics of the G.711 voice codec without Packet Loss Concealment (PLC) [8].
The payload size is set to 160 bytes and the sending interval to 20 ms in G.711
codec without PLC. The resulting throughput of VoIP traffic is 64 Kbps. We collected 10 300-second-long data sets after transmitting voice packets between the
MN and the CN. Because the clocks on the MN and CN were not synchronized,
we could not measure the one-way delay accurately. Instead, we took round-trip
measurements of VoIP traffic, and halved the delay. Due to the difference in
uplink and downlink bandwidths, half the round-trip delay is likely to be larger
than the one-way uplink delay. However, the WiBro link was very lightly loaded

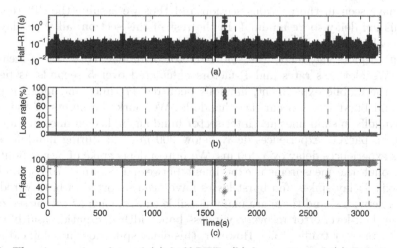

Fig. 4. The time-series plots of (a) half-RTT, (b) loss rate, and (c) R-factor. Each
vertical line indicates the time when either inter-RAS(solid line) or inter-sector(dashed
line) handoff occurs.

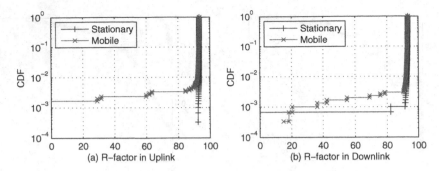

Fig. 5. CDF of R-factor in uplink and downlink

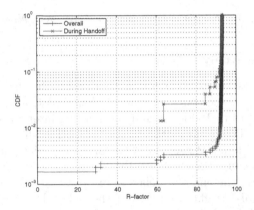

Fig. 6. CDF of R-factor in 5-sec-intervals vs. during handoffs

as we have seen in the previous section and thus we assume the difference in transmission delay to be minimal. In the rest of this section, all the delays we use are calculated as described.

Figure 4 plots the delay, loss, and R-factor of mobile VoIP traffic measurements. We plot loss rates and R-factors calculated over 5 seconds as before. During the mobile experiment, handoffs occurred 17 times and most of delay spikes and burst losses occur near handoffs. We mark the points of the inter-RAS handoffs in solid lines and inter-sector handoffs in dashed lines in Figure 4. Almost all packets experiences delay below 200 ms, but during handoffs some packets experiences delay over 400 ms. We note a delay spike of 5 sec about half way through the measurement experiment between 1600 and 1700 seconds on the x-axis. Delay spikes and burst losses have been reported in both wired and wireless networks, and there are many possible causes, such as cell or sector reselection, link-level error recovery, wireless bandwidth fluctuation and blocking by higher priority traffic [15,5]. However, this delay spike does not coincide with a handoff, and needs further investigation.

Figure 5 shows the R-factor of VoIP quality of uplink (MN to CN) and downlink (CN to MN) in the stationary and mobile cases. The R-factor is calculated every 5 seconds. From the figure we have found that more than 99% of R-factors are above 90 in both the stationary and the mobile case and only 0.2% of R-factors is below 70 in the mobile case. The R-factor of 70 or above is considered toll-quality, and thus mobile devices attached to the WiBro network are likely to experience toll-quality using VoIP applications.

To quantify the impact of handoffs on QoS of VoIP, we have computed R-factors using average delay and loss rate for the interval of 5 seconds before and after the handoff and compared with the overall cumulative distribution function (CDF) of R-factors in Figure 6. Here again we observe that the about 99% of R-factors during handoffs are more than 85, which translates to good quality for voice communication.

5   Summary and Future Work

In this work, we have conducted experiments to evaluate QoS of VoIP applications over the WiBro network. In order to capture the baseline performance of the WiBro network we measure and analyze the characteristics of delay and throughput under stationary and mobile scenarios. Then we evaluate QoS of VoIP applications using the E–Model of ITU–T G.107. Our measurements show that the achievable maximum throughputs are 5.3 Mbps in downlink and 2 Mbps in uplink. VoIP quality is better than or at least as good as toll quality despite user mobility exceeding the projected limit of WiBro mobility support. Using RAS and sector identification information, we show that the handoff is correlated with throughput and quality degradation.

The WiBro network is in its early phase of deployment and about 70,000 subscribers are reported to have signed up. As our CBR traffic analysis shows, the network is very lightly loaded, allowing near maximum throughputs for many 5-second-long periods. In future, we plan to conduct more experiments with cross-traffic injection and TCP traffic.

Acknowledgement

Sue Moon, Keon Jang, and Dooyoung Lee were supported by KOSEF through AITrc (Advanced Information Technology Research Center). Youngseok Lee was supported by the Ministry of Information and Communication, Korea, under the ITRC (Information Technology Research Center) support program supervised by the IITA (Institute of Information Technology Advancement) (IITA-2008-(C1090-0801-0016)).

References

1. Cicconetti, C., Lenzini, L., Mingozzi, E., Eklund, C.: Quality of service support in IEEE 802.16 networks. IEEE Network 20, 50–55 (2006)
2. Cole, R.G., Rosenbluth, J.H.: Voice over IP performance monitoring. SIGCOMM Comput. Commun. Rev. 31(2), 9–24 (2001)

3. D-ITG, http://www.grid.unina.it/software/itg/
4. Ghosh, A., Wolter, D.R., Andrews, J.G., Chen, R.: Broadband wireless access with WiMax/802.16: current performance benchmarks and future potential. IEEE Communications Magazine 43, 129–136 (2005)
5. Hornero, R., Abasolo, D.E., Jimeno, N., Espino, P.: Applying approximate entropy and central tendency measure to analyze time series generated by schizophrenic patients. In: Proc. of the 25th Annual International Conference of the IEEE, September 2003, pp. 2447–2450 (2003)
6. iperf, http://dast.nlanr.net/projects/iperf/
7. ITU–T standard G.107. The e-model, a computational model for use in transmission planning (March 2005)
8. ITU–T standard G.711. Pulse code modulation (PCM) of voice frequencies (November 1988)
9. ITU–T standard P.800. Methods for subjective determination of transmission quality (August 1996)
10. ITU–T standard P.861. Objective quality measurement of telephoneband (300–3400 hz) speech codecs (August 1996)
11. ITU–T standard P.862. Perceptual evaluation of speech quality (pesq): An objective method for end-to-end speech quality assessment of narrow-band telephone networks and speech codecs (February 2001)
12. Lee, H., Kwon, T., Cho, D.-H., Lim, G., Chang, Y.: Performance analysis of scheduling algorithms for VoIP services in IEEE 802.16e systems. In: Proc. of Vehicular Technology Conference, vol. 3, pp. 1231–1235 (2006)
13. TTAS.KO-06.0065R1. WiBro standard phase–I (December 2004)
14. WiMax, http://en.wikipedia.org/wiki/wimax
15. Xin, F., Jamalipour, A.: TCP performance in wireless networks with delay spike and different initial congestion window sizes. Computer Communications 29, 926–933 (2006)

Packet Sampling for Flow Accounting: Challenges and Limitations

Tanja Zseby[1], Thomas Hirsch[1], and Benoit Claise[2]

[1] Fraunhofer Institute FOKUS, Kaiserin-Augusta-Allee 31, 10589 Berlin, Germany
{Tanja.Zseby,Thomas.Hirsch}@fokus.fraunhofer.de
[2] Cisco Systems, De Kleetlaan 6a b, 1831 Diegem, Belgium
bclaise@cisco.com

Abstract. We investigate the applicability of packet sampling techniques to flow-based accounting. First we show by theoretical considerations how the achievable accuracy depends on sampling techniques, parameters and traffic characteristics. Then we investigate empirically which accuracy is achieved with typical flow characteristics by experiments with real traffic traces from three different networks. In a third step we illustrate how to support sampling-based accounting by providing an accuracy statement together with the measured data. We show which information is required for this and how an accuracy assessment can be approximated from information available after the sampling process using information elements of the IP flow information export protocol (IPFIX).

Keywords: packet sampling, accounting, IPFIX.

1 Introduction

Sampling aims at the reduction of measurement costs by estimating the metric of interest from a subset of data. It is important that the extent of potential estimation errors can be evaluated, especially if measurement results map to monetary values as it is the case for accounting. The achievable accuracy usually depends on characteristics of the population, i.e., in our case the traffic in the network. Since network traffic is extremely dynamic providing an up-to-date accuracy assessment is not trivial. It must be derived from the limited information available after the sampling process. It has to be calculated per flow and updated continuously.

Basic packet selection methods are currently standardized in the IETF PSAMP group [6]. A flow sampling scheme for accounting is introduced in [1]. Sample and Hold [2], Shared-state Sampling (S3) [3], and the Runs bAsed Traffic Estimator (RATE) [4] propose packet sampling methods that bias the selection process towards large flows in order to reduce resource consumption for flow caching and flow record transfer. This makes sense for accounting because in typical flow distributions a few large flows contribute to the majority to the overall traffic volume (e.g. [1]). Nevertheless, all those approaches require the classification of packets into flows before or during the sampling process. In contrast to this we investigate the effects of packet sampling that is applied *before* flow classification, so that only selected

M. Claypool and S. Uhlig (Eds.): PAM 2008, LNCS 4979, pp. 61–71, 2008.

packets need to be classified, which significantly reduces workload on routers [5]. We compare the achievable accuracy for basic PSAMP schemes and a stratified method used in Cisco NetFlow to accounting requirements. We show how the accuracy can be approximated from available information, using IPFIX information elements [11].

2 Flow Accounting Requirements

The accuracy of an estimate is assessed by **bias** and **precision**. For accounting we should only use unbiased estimates. This is the case if the expectation of the estimated values equals the real value. The precision is derived from the variance (or its square root: the **standard error**) of the estimate and expresses how far estimated values from sample runs would spread. The higher the standard error the lower is the precision. An accuracy statement can be presented to customers by a **confidence interval**. Confidence boundaries define the area in which the real value should lie and can be expressed by the maximum tolerable estimation error. The confidence level (CL) gives the probability that the real value lies within this range. From this we can derive a maximum standard error that should not be exceeded if a given accuracy is required. Table 1 shows the maximum relative standard error for different accuracy requirements for a normal distributed estimate.

Table 1. Maximum Relative Standard Error for Different Accuracy Requirements

Rel. Est. Error	CL	Rel. StdErr	Rel. Est. Error	CL	Rel. StdErr
0.01 (1%)	99%	0.003876	0.1 (10%)	95%	0.051020
0.01 (1%)	95%	0.005102	0.15 (15%)	95%	0.076531
0.05 (5%)	99%	0.019380	0.20 (20%)	95%	0.102041
0.05 (5%)	95%	0.025510	0.30 (30%)	95%	0.1531

3 Accuracy Assessment in Theory

We here provide a theoretical assessment of bias and precision by providing formulas for expectation and standard error for the sampling schemes. We also give formulas for sampling after classification, but our focus is on sampling before classification. It is the more complex case, saves classification effort and is used in NetFlow.

Accuracy Assessment for n-out-of-N Sampling. In *n-out-of-N* sampling exactly n elements are selected from the population, which consists of N elements [6]. If there is only one flow ($N=N_f$) in the traffic mix or we apply sampling after classification, the number N_f of packets per flow is known. The number n_f of selected packets can be set per flow and is also known.

The estimate $\hat{Sum}_f$ for the number of bytes in flow f can be simply calculated from the packet sizes $x_{i,f}$ of the selected packets, by extrapolating with n_f and N_f. The expected bias is zero. The standard error can be calculated by the standard formula for an n-out-of-N selection [9] from sampling parameters and packet size variance $\sigma^2_{x_f}$.

$$Sûm_f = \frac{N_f}{n_f} \cdot \sum_{i=1}^{n_f} x_{i,f} \tag{1}$$

$$StdErr_{abs}\left[Sûm_f\right] = N_f \cdot \frac{\sigma_{x_f}}{\sqrt{n_f}} \cdot \sqrt{\frac{N_f - n_f}{N_f - 1}} \tag{2}$$

If we apply sampling before classification N_f and n_f are unknown. Extrapolation must be done with the overall population N and sample size n.

$$Sûm_f = \frac{N}{n} \cdot \sum_{i=1}^{n_f} x_{i,f} \tag{3}$$

In contrast to the case above (where $n_f = n = const$), here the number n_f of packets from flow f in the sample varies for each sampling run and has to be considered as random variable (r.v.) itself. The estimate contains two random variables, n_f and $x_{i,f}$. To assess the estimation quality we need to calculate expectation and variance of a sum of random variables, where the number of addends itself is a random variable. We model n_f as a discrete r.v. with a binomial distribution[1] $B(n, N_f/N)$. We denote the mean packet size of all packet sizes in flow f in the population by μ_{x_f} and their variance by $\sigma_{x_f}^2$. With $x_{i,f}$ we denote the number of bytes of the ith selected packet[2]. Since we apply a random selection, the $x_{i,f}$ are independent identical distributed (i.i.d.).

With the assumption of the binomial distribution for n_f and independency for the $x_{i,f}$ we can derive the following formulas for expectation and variance for the estimated sum for flow f (see appendix):

$$E\left[Sûm_f\right] = N_f \cdot \mu_{x_f} = Sum_f \tag{4}$$

$$V\left[Sûm_f\right] = \frac{1}{n} \cdot \left(N \cdot N_f \cdot \left(\sigma_{x_f}^2 + \mu_{x_f}^2\right) - N_f^2 \cdot \mu_{x_f}^2\right) \tag{5}$$

The expectation equals the real volume, i.e. the estimation is unbiased. The variance of the estimated flow volume, and with this the expected accuracy of the estimation depends on the parameters n, N, N_f, μ_{x_f} and $\sigma_{x_f}^2$. Sample size n and population size N are preconfigured sampling parameters. N_f, μ_{x_f} and $\sigma_{x_f}^2$ are flow characteristics. N_f denotes the number of packets in the population that belong to flow f. The packet size mean μ_{x_f} and the packet size variance $\sigma_{x_f}^2$ depend on the packet size distribution in flow f. If we take the square root of the variance we get the absolute standard error.

$$StdErr_{abs}\left[Sûm_f\right] = \sqrt{\frac{1}{n} \cdot \left(N \cdot N_f \cdot \left(\sigma_{x_f}^2 + \mu_{x_f}^2\right) - N_f^2 \cdot \mu_{x_f}^2\right)} \tag{6}$$

A division by the flow volume provides the relative standard error (see appendix).

---

[1] If $f \leq 0.05$ and $0.1 < N_f/N < 0.9$ the hyper geometrical distribution $Hy(N, N_f, n)$ can be approximated by a binomial distribution $B(n, N_f/N)$ (see e.g., [10]).

[2] Note that the index i is used for the selected packets only and not for all packets in the flow.

Accuracy Assessment for 1-in-K Sampling (stratified). Cisco NetFlow implements a sampling scheme that we call 1-in-K sampling[3]. 1-in-K sampling is a count-based stratified n-out-of-N sampling. The selection process is done in two steps. First the measurement interval is divided into L subintervals of size K. Then one packet is randomly selected per subinterval. The measurement interval, i.e., the population for which a parameter should be estimated, still consists of N packets. The estimate is calculated from all n_f packets that were selected in all subintervals in the measurement interval.

$$Sûm_f = \frac{N}{n} \cdot \sum_{i=1}^{n_f} x_{i,f} \text{ with } n_f = k_{f,1} + k_{f,2} + \ldots + k_{f,L} \tag{7}$$

The difference to n-out-of-N sampling is that here the number n_f of packets from flow f in the sample does not necessarily follow a binomial distribution. The sample size k within the subinterval is always 1. The number k_f of packets from flow f within this sample can be 0 or 1. The probability that k_f is 1 (i.e., the selected packet belongs to flow f) depends on the total amount of packets from flow f in the subinterval K_f. Therefore k_f can be considered as a Bernoulli distributed random variable with a probability of success $p_f = K_f/K$. So the distribution of n_f depends on those subinterval probabilities, which depend on the packets per flow in the subinterval.

If all packets in the measurement interval belong to one flow ($N_f = N$), the standard error for stratified sampling can be calculated as follows [see [9], following equation 5.9]:

$$StdErr[Sûm]_{strat} = \sqrt{\sum_{l=1}^{L} K_l \cdot (K_l - k_l) \cdot \frac{\sigma_{x,l}^2}{k_l}} \tag{8}$$

In the 1-in-K sampling implemented in NetFlow all strata have the same size ($K_l = N/L$) and only one packet is selected per stratum ($k_l = 1$). Furthermore, if $K_l >> k_l$ we can approximate $K_l - k_l \approx K_l$. With this we get

$$StdErr[Sûm]_{strat} = \sqrt{\sum_{l=1}^{L} K_l^2 \cdot \frac{\sigma_{x,l}^2}{k_l}} = \sqrt{\frac{N^2}{L^2} \cdot \sum_{l=1}^{L} \sigma_{x,l}^2} = N \cdot \sqrt{\frac{1}{L^2} \cdot \sum_{l=1}^{L} \sigma_{x,l}^2} \tag{9}$$

The accuracy depends on the number L of strata and on the packet size variances $\sigma_{x,l}^2$ in the subintervals.

If the packets in the measurement interval belong to different flows ($N_f < N$), one has to consider not only the distribution of packet sizes over the subintervals but also the distribution of flow IDs. The calculation of the standard error becomes more complex because the variances have to be calculated per strata. The standard error now depends on the per-flow characteristics (number of packets K_f, packet size variance $\sigma_{x_f,l}^2$, and mean $\mu_{x_f,l}$) within each subinterval.

$$StdErr[Sûm_f]_{strat} = \sqrt{\sum_{l=1}^{L} \left(K_{f,l} \cdot K \cdot \left(\sigma_{x_f,l}^2 + \mu_{x_f,l}^2 \right) - \mu_{x_f,l}^2 \cdot K_{f,l}^2 \right)} \tag{10}$$

---

[3] To avoid confusion with the interval length N we call the scheme 1-in-K instead of 1-in-N.

The vigilant reader may miss the sampling parameters n and N in the formula. But for 1-in-K sampling the population size N is formed by the stratum size K and the number of strata L ($N=K*L$). The sample size n equals the number of strata L.

Theoretical Comparison of Schemes. A scheme provides a higher estimation accuracy if the standard error is smaller. That means 1-in-K sampling performs better if the following condition holds:

$$StdErr[Sûm]_{strat} < StdErr[Sûm]_{rand} \tag{11}$$

If we consider only one flow a stratification gain can be achieved if:

$$N \cdot \sqrt{\frac{1}{L^2} \cdot \sum_{l=1}^{L} \sigma_{x,l}^2} < N \cdot \sqrt{\frac{\sigma_x^2}{n}} \tag{12}$$

Since $n=L$, this can be simplified to.

$$\frac{1}{L} \cdot \sum_{l=1}^{L} \sigma_{x,l}^2 < \sigma_x^2 \tag{13}$$

That means we get a higher accuracy with 1-in-K sampling if the mean of the variances per subinterval (over all subintervals) is smaller than the variance within the whole measurement interval.

For multiple flows the formula gets more complex, because per-flow characteristics need to be taken into account. With the formulas for the standard error for n-out-of-N and stratified sampling for case II we get:

$$\sqrt{L \sum_{l=1}^{L} \left(K_{f,l} K \left(\sigma_{x_f,l}^2 + \mu_{x_f,l}^2 \right) - \mu_{x_f,l}^2 K_{f,l}^2 \right)} < \sqrt{\frac{1}{n} \left(N N_f \left(\sigma_{x_f}^2 + \mu_{x_f}^2 \right) - N_f^2 \mu_{x_f}^2 \right)} \tag{14}$$

In order to assess the accuracy for 1-in-K sampling one would need information about per flow characteristics for each subinterval. In contrast to n-out-of-N sampling those parameters cannot be approximated for 1-in-K sampling.

4 Accuracy Assessment in Practice

As we have seen we need the flow characteristics to calculate the accuracy. Since those are unknown, they have to be estimated from sampled values. A second problem is the amount of data that needs to be stored to provide an accuracy statement. Storing per-packet information results in too much data even if only sampled packets are stored. Therefore we here show how to calculate the accuracy from aggregated information. In addition we show how IPFIX Information Elements (IEs) can be utilized to export the required values needed for the accuracy assessment.

Accuracy Assessment from Sampled Packets. With the sampling parameters, the number of the sampled packets and their packet sizes we can provide estimates for the relevant parameters for n-out-of-N sampling.

$$\hat{N}_f = \frac{N}{n} \cdot n_f \ (15) \quad \hat{\mu}_{x_f} = \overline{x}_f = \frac{1}{n_f} \cdot \sum_{i=1}^{n_f} x_{i,f} \ (16) \quad \hat{\sigma}^2_{x_f} = s^2_{x_f} = \frac{1}{n_f - 1} \cdot \sum_{i=1}^{n_f} (x_{i,f} - \overline{x}_f)^2 \ (17)$$

Using those estimates in formula (5) results in the following equation:

$$\hat{V}[\hat{Sum}_f] = \frac{N^2}{n} \cdot \left(\frac{n_f}{n} \cdot \left(s^2_{x_f} + \overline{x}^2_{x_f} \right) - \overline{x}^2_{x_f} \cdot \frac{n_f^2}{n^2} \right) \tag{18}$$

For 1-in-K sampling the assessment from sampled values is problematic. As can be seen from the formulas in section 3 we would need to estimate $K_{f,l}$, $\mu_{f,l}$ and $\sigma^2_{f,l}$ per subinterval. Since we select only one packet per subinterval, it is not possible to calculate acceptable estimates for mean and variance. As a consequence we cannot provide a practical accuracy statement from the sampled values for 1-in-K sampling. In empirical investigations we have seen that for many flows the accuracy for 1-in-K is close to the n-out-of-N model with current packet size distributions. Therefore the n-out-of-N accuracy often provides a good approximation.

Accuracy Assessment from Aggregated Information and IPFIX. Cisco currently stores for each flow the number n_f of packets in the sample and the sum of packet sizes from the sampled packets. With these two values and the sampling parameters n and N, one can easily calculate the estimates $\hat{N}_f$ and $\overline{x}_f$ ((15),(16)). But the calculation of the estimated variance $s^2_{x_f}$ is not possible with the stored values. A calculation of $s^2_{x_f}$ using (17) would require knowledge about all packet sizes in the sample. In order to avoid the storage of all packet sizes from the sampled packets, one can use an alternative variance calculation based on the sum and the square sum of the selected packet sizes.

$$s^2_{x_f} = \frac{1}{n_f - 1} \cdot \sum_{i=1}^{n_f} x^2_{i,f} - \frac{1}{n_f \cdot (n_f - 1)} \cdot \left(\sum_{i=1}^{n_f} x_{i,f} \right)^2 \tag{19}$$

Sum and square sum can be updated when a packet is selected and the packet sizes themselves do not need to be stored. If we insert (19) into formula (18) one can easily derive the accuracy from the stored aggregated values (sum and square sum). We recommended the storage of the square sum to Cisco. It has been added as an information element to the flow information export protocol IPFIX [12], and therefore will be available in Cisco routers in future. Table 2 shows the IPFIX and PSAMP information elements ([11], [13]) that provide the required values for calculating an accuracy statement.

If sampling is applied those values are calculate from the sampled packets and can be used to derive the required estimates. For count-based measurement intervals the number of packets in the measurement interval is preconfigured and can be reported with the samplingPopulation IE. For time-based measurement intervals one can report the number by defining an IPFIX flow that comprises all packets on the link

Table 2. IPFIX/PSAMP Information Elements

Parameter	IPFIX/PSAMP IEs
Number N of packets in measurement interval	samplingPopulation
Number n of packets in sample	samplingSize
Number of packets from flow f in sample	packetTotalCount
Sum (bytes in sampled packets)	octetTotalCount
Square sum (bytes in sampled packets)	octetTotalSumOfSquares

and use the packetTotalCount information element for this flow. An alternative is to use link packets counters from SNMP.

5 Experiments

We investigate the achievable accuracy for different schemes, classification rules and interval lengths with real traffic traces from 3 different networks. We show how many flows conform to given accuracy requirements.

Traces. The first trace set is from a large European operator (denoted as OP). The second set we collected at CIRIL [17], a regional network provider that interconnects universities and research institutes with the French Research and Education Network RENATER. Measurements were taken on a 1 Gbit multimode Ethernet access link to the national research network. As a third source we used the 6 hour traces *NZIX07m06d12h* (NZIX1) and *NZIX07m06d06h* (NZIX2) from [14]. We performed experiments with two different classification schemes. S24D24 distinguishes flows with respect to source and destination network both with a 24 bit netmask. S24D00 distinguishes flows only with respect to the source network. If packets of the same flow are observed in different measurement intervals they are counted as separate flows. Table 3 shows the number of flows observed for different classification rules and interval lengths (in number of packets). We use a letter per setting as identifier.

Table 3. Trace Characteristics

Setting	Trace	Size	#packets	Classification	MI	#flows
A	OP1	15 GB	122,800,288	S24D00	10M	852,593
B	OP1	15 GB	122,800,288	S24D24	10M	5,354,933
C	OP2	92 GB	766,071,712	S24D00	10M	69,001
D	CIRIL	2 GB	34,324,092	S24D00	10M	3,588,520
E	NZIX1	2 GB	65672186	S24D00	10M	8,569
F	NZIX2	39 GB	770,842,909	S24D00	10M	4,093
G	NZIX1	2 GB	65672186	S24D00	1M	79,383
H	NZIX1	2 GB	65672186	S24D24	1M	53,7138

Fig. 1 (left) shows a summarized representation of all flows in the CIRIL trace (setting D). Each dot represents a flow. The dimensions are the three flow characteristics that are relevant for the estimation accuracy: number of packets, packet size mean and variance (represented by the standard deviation). With settings D the trace contains 3,588,520 flows. The majority of flows are small. Only 4,624 flows consist of more than 200,000 packets (not shown in graph). The peak at the standard deviation of zero and small means is caused by flows with packets of equal sizes. Several flows consist of only one packet. Those also have a standard deviation of zero. For the other traces and settings we observed similar flow distributions. Especially the existence of a majority of small flows was observed for all traces.

Conformance to Accuracy Requirements. First we calculate the achievable accuracy using the observed real flow characteristics and formula (6). Table 4 shows how many flows in the traces conform to given accuracy requirements for a sampling fraction of $f=5\%$. The accuracy is given by the threshold t for the standard error.

Table 4. Conformant Flows for n-out-of-N, $f=5\%$

ID	Number of Conformant Flows for rel. StdErr $\leq$ t				
	$t=0.003876$	$t=0.005102$	$t=0.019380$	$t=0.025510$	$t=0.051020$
A	0	1	1330	3,316	25,746
B	0	0	8	38	659
C	2	5	30	66	310
D	300	578	12,475	19,984	56,904
E	0	0	63	98	425
F	7	21	276	437	1,414
G	0	0	64	72	421
H	0	0	0	0	311

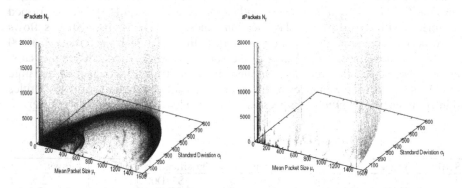

Fig. 1. Setting D: All Flows (left) and Conformant Flows (right)

Common accuracy requirements for accounting are a maximum relative error of *0.01* or *0.05* with a confidence level of at least *95%*. With a sampling fraction of *5%* the achievable accuracy is too low for the vast majority of flows for all settings. Fig. 1 (right) shows the flows conformant to *StdErr* $\leq$ *0.05*. Only flows with a large number of packets N_f achieve an acceptable accuracy.

Flow Conformance from Empirical Tests. In a second step we investigate the standard error empirically from sampling tests. For this we performed $R=1,000$ sampling runs for each scheme. Table 5 shows the results from experiments with setting G and different schemes.

Table 5. Conformant Flows for Setting G (*NZIX1, S24D00, f=5%*)

Max rel. StdErr	Error/CL	n-of-N	1-in-K	Systematic
0.003876	0.01/99%	0	0	0
0.005102	0.01/95%	0	0	0
0.019380	0.05/99%	64	64	62
0.025510	0.05/95%	72	72	83
0.051020	0.1/95%	473	475	567
0.076531	0.15/95%	1406	1425	1580
0.102041	0.2/95%	2316	2568	2860
0.1531	0.3/95%	5146	5397	5799
>0.1531	-	79383	79383	79383

The numbers for n-out-of-N sampling correspond quite well to those derived from the formula Table 4. For 1-in-K sampling we get quite similar numbers. This is in line with previous tests we performed about the scheme differences. Systematic sampling performs a little bit better, but the standard errors in the tests differed much from those of n-out-of-N. A theoretical prediction is problematic. Again, only few flows get accuracies sufficient for accounting. In order to achieve higher accuracies per flow one can increase the sample fraction, work with more coarse grained classifications or modify the measurement interval length. When modifying the measurement interval length it is relevant how flow characteristics evolve in order to assess the accuracy (see section 3).

6 Conclusion

We investigated the applicability of packet sampling to flow accounting. We analyzed basic PSAMP schemes and a stratified scheme used in Cisco NetFlow and showed how the accuracy depends on flow parameters and measurement settings. Theoretical considerations were supplemented by experiments with traffic traces from three different networks. The accuracy for sampling before classification was very poor. The main reason is the high number of small flows in the traces. Longer observation periods, coarse grained classification or the aggregation of flows results in larger flows and higher accuracies. A further option is to use a biased flow selection based on the expected accuracy. In addition we showed how the accuracy can be derived from sampled values and aggregated information stored in routers during run-time. For this, Cisco has included the storage of the square sum of the packet sizes in NetFlow.

References

[1] Duffield, N., Lund, C., Thorup, M.: Charging from Sampled Network Usage. In: ACM Internet Measurement Workshop IMW 2001, San Francisco, USA, November 1-2 (2001)
[2] Estan, C., Varghese, G.: New Directions in Traffic Measurement and Accounting: Focusing on the Elephants, Ignoring the Mice. ACM Transactions on Computer Systems (August 2003)

[3] Raspall, F., Sallent, S., Yufera, J.: Shared-state sampling. In: Proceedings of the 6th Internet Measurement Conference (IMC 2006), Rio de Janeiro, Brazil (2006)

[4] Kodialam, M., Lakshman, T.V., Mohanty, S.: Runs bAsed Traffic Estimator (RATE): A Simple, Memory Efficient Scheme for Per-Flow Rate Estimation. In: IEEE INFOCOM 2004, Hong Kong (2004)

[5] NetFlow Performance Analysis, Cisco white paper (2005),
http://www.cisco.com/en/US/products/ps6601/
products_white_paper0900aecd802a0eb9.shtml

[6] Zseby, T., Molina, M., Duffield, N., Niccolini, S., Raspall, F.: Sampling and Filtering Techniques for IP Packet Selection. Internet Draft <draft-ietf-psamp-sample-tech-10.txt> (work in progress, June 2007)

[7] Quittek, J., Zseby, T., Claise, B., Zander, S.: Requirements for IP Flow Information Export (IPFIX). In: RFC 3917 (October 2004)

[8] Zseby, T.: Stratification Strategies for Sampling-based Non-intrusive Measurements of One-way Delay. In: Proceedings of Passive and Active Measurement Workshop (PAM 2003) April 6-8 (2003)

[9] Cochran, W.G.: Stichprobenverfahren. Walter de Gruyter &Co, Berlin, New York (1972)

[10] Schwarz, H.: Stichprobenverfahren. Oldenbourg Verlag, GmbH (1975)

[11] Quittek, J., Bryant, S., Claise, B., Aitken, P., Meyer, J.: Information Model for IP Flow Information Export. In: RFC 5102 (January 2008)

[12] Claise, B. (ed.): Specification of the IP Flow Information Export (IPFIX) Protocol for the Exchange of IP Traffic Flow Information. In: RFC 5101 (January 2008)

[13] Dietz, T., Dressler, F., Carle, G., Claise, B., Aitken, P.: Information Model for Packet Sampling Exports, Internet-Draft draft-ietf-psamp-info-07.txt (work in progress, October 2007)

[14] Waikato Internet Traffic Storage (WITS),
http://wand.cs.waikato.ac.nz/wand/wits/

[15] Fisz, M.: Probability Theory and Mathematical Statistics, 3rd edn. Robert E. Krieger Publishing Company Inc, Malabar, Florida (1963)

[16] Wentzel, E.S., Owtscharow, L.A.: Aufgabensammlung zur Wahrscheinlichkeitsrechnung. Akademieverlag, Berlin (1975)

[17] Centre Interuniversitaire de Ressources Informatiques de Lorraine (CIRIL),
http://www.ciril.fr/

Appendix: Expectation and Variance for n-out-of-N Sampling

The random variable $x_{i,f}$ denotes the packet size of the ith selected packet from flow f. Since a random selection is applied, we can assume that the $x_{i,f}$ are statistically independent. Since n_f follows a binomial distribution, the expectation and variance of n_f is given by formulas for a binomial distribution:

$$E[n_f] = n \cdot \frac{N_f}{N} \quad (20) \qquad V[n_f] = n \cdot \frac{N_f}{N} \cdot \left(1 - \frac{N_f}{N}\right) \quad (21)$$

With these considerations, the task is reduced to the calculation of expectation and variance of a r.v. Z, where Z is the sum of independent identical distributed (i.i.d.) random variables X and the number of summands Y is a binomial distributed random variable. The expectation of such a r.v. is given in [15].

$$E[Z] = E[X] \cdot E[Y] \quad \text{for} \quad Z = \sum_{i=1}^{Y} X_i \quad (22)$$

With this the expectation of the estimated volume is calculated as follows:

$$E[Z] = \frac{N}{n} \cdot E\left[\sum_{i=1}^{n_f} x_{i,f}\right] = \frac{N}{n} \cdot E[x_{i,f}] \cdot E[n_f] = \frac{N}{n} \cdot \mu_{x_f} \cdot n \cdot \frac{N_f}{N} = N_f \cdot \mu_{x_f} = Sum_f \quad (23)$$

The expectation of the estimate equals the real volume, i.e. the estimation is unbiased. A formula to calculate the variance for this special case, but for continuous random variables is derived in [16]. This formula can be also applied for discrete variables.

$$V[Z] = E[Y] \cdot V[X] + E[X]^2 \cdot V[Y] \quad \text{for} \quad Z = \sum_{i=1}^{Y} X_i \quad (24)$$

With this the variance of the estimated flow volume can be expressed as follows:

$$V\left[S\hat{u}m_f\right] = \frac{N^2}{n^2} \cdot V\left[\sum_{i=1}^{n_f} x_{i,f}\right] = \frac{N^2}{n^2} \cdot \left(E[n_f] \cdot V[x_{i,f}] + E[x_{i,f}]^2 \cdot V[n_f]\right)$$

$$= \frac{N^2}{n^2} \cdot \left(E[n_f] \cdot V[x_{i,f}] + E[x_{i,f}]^2 \cdot V[n_f]\right) \quad (25)$$

The relative standard error can be easily derived from the variance.

$$StdErr_{rel}\left[S\hat{u}m_f\right] = \frac{StdErr_{abs}\left[S\hat{u}m_f\right]}{Sum_f} = \frac{\sqrt{\frac{1}{n} \cdot \left(N \cdot N_f \cdot \left(\sigma_{x_f}^2 + \mu_{x_f}^2\right) - N_f^2 \cdot \mu_{x_f}^2\right)}}{N_f \cdot \mu_{x_f}} \quad (26)$$

On the Validation of Traffic Classification Algorithms

Géza Szabó, Dániel Orincsay, Szabolcs Malomsoky, and István Szabó

TrafficLab, Ericsson Research, Budapest, Hungary
{geza.szabo,daniel.orincsay,szabolcs.malomsoky,istvan.szabo}@ericsson.com

Abstract. Detailed knowledge of the traffic mixture is essential for network operators and administrators, as it is a key input for numerous network management activities. Traffic classification aims at identifying the traffic mixture in the network. Several different classification approaches can be found in the literature. However, the validation of these methods is weak and ad hoc, because neither a reliable and widely accepted validation technique nor reference packet traces with well-defined content are available. In this paper, a novel validation method is proposed for characterizing the accuracy and completeness of traffic classification algorithms. The main advantages of the new method are that it is based on realistic traffic mixtures, and it enables a highly automated and reliable validation of traffic classification. As a proof-of-concept, it is examined how a state-of-the-art traffic classification method performs for the most common application types.

1 Introduction

The aim of traffic classification is to find out what type of applications are run by the end users, and what is the share of the traffic generated by the different applications in the total traffic mix. Research for better and better traffic classification methods is blooming with the constant increase of network capacity, the emerging application types, and common usage of traffic deceiving techniques. However, the objective comparison of these methods has not been possible yet due to several reasons. Firstly, there are no perfectly classified traffic traces available. Moreover, the validation is typically done with another specific classification method. This situation results in such anarchy that papers can state nearly anything about their introduced method as there is no chance to check it by others or verify with a commonly known and accepted reference test.

In this paper we provide a validation method, which can reliably test the accuracy of traffic classification algorithms. In practice, the objective is typically to identify applications in passively observed traffic. We believe that such a classification method can be convincingly validated only by an active test, for which a number of requirements are fulfilled, such as:

- It should be independent from classification methods, i.e. the validation of a classification method by another one must be avoided,

M. Claypool and S. Uhlig (Eds.): PAM 2008, LNCS 4979, pp. 72–81, 2008.

- About each packet the test should provide reference information that can be compared to the result of the classification method under study,
- The test should be deterministic, meaning that it should not rely on any probabilistic decisions,
- Feasibility: it should be possible to create large tests in a highly automated way, and
- The environment where the active measurements are collected should be realistic.

The paper is organized as follows: in Section 2 an overview of existing traffic classification methods is provided together with a discussion of the techniques and datasets used to validate them. In Section 3 a new method is introduced which makes it possible to validate traffic classification methods. In Section 4, a state-of-the-art traffic classification method is validated as a proof-of-concept, demonstrating how it performs for several application types that are included in the example test.

2 Existing Traffic Classification Methods and Their Evaluation

Currently, there are a couple of fundamentally different approaches for traffic classification. In this section we browse through the state-of-the-art traffic classification methods. We discuss briefly the accuracy of these methods, which is relevant here, because in most cases a classification method is validated by another classification method ([12], [19], [18]).

The most accurate traffic classification would obviously be complete protocol parsing. However, many protocols are ciphered due to security reasons (SSH [5], SSL [4]). Also some are proprietary, thus there is no public description available (Skype [6], MSN Messenger [2], World of Warcraft [9], etc.). In general, it would be difficult to implement every protocol which can occur in the network. In addition, even simple protocol state tracking can make the method so resource consuming that it becomes practically infeasible.

- **Port based classification:** In the simplest and most common method the classification is based on associating a well-known port number with a given traffic type, e.g., web traffic with TCP port 80 [1]. This method needs access only to the header of the packets. The port based method becomes insufficient in many cases, since no specific application can be associated to a dynamically allocated port number, or the traffic classified as web may easily be something else tunneled via HTTP. The port based method is a standard, common method, however due to the above problems, it can not be considered to be reliable.
- **Signature based classification:** To make protocol recognition feasible, only specific byte patterns are searched in the packets in a stateless manner. These byte signatures are predefined to make it possible to identify particular traffic types, e.g., web traffic contains the string 'GET', eDonkey P2P

traffic contains 'xe3x38'. The common feature of the signature based heuristic methods is that in addition to the packet header, they also need access to the payload of the packets. Especially in the case of well documented open protocols, this method can work well. However, in practice only extensive experiences with real traces provide enough feedback to select the best performing byte signatures. For example, the 'GET' message could be the criterion of both HTTP and Gnutella (a P2P protocol), thus this signature alone, without applying other criteria, is not proper for accurate traffic classification. The main disadvantage of the signature based method is that the signatures have to be kept up to date, otherwise some applications can be missed, or the method can produce false positives. The other disadvantage is that this method cannot deal with encrypted content.

Authors of [16] validated their constructed signature database by manually checking the false positive ratio of their technique. Their approach was to investigate TCP connections which were identified as P2P connections. If in fact the content of the connection did not belong to a P2P protocol they counted the connection as a false positive. By the term 'active measurements' they mean that specific traffic type is generated on purpose, thus what kind of traffic is expected can be exactly known at a certain point in the measurement. This is the most common way of developing signature databases as this method ensures that the traffic is sterile, i.e., only a specific application is measured at a time. The measurements they used are not public, therefore others cannot use them as reference.

– **Connection pattern based classification:** The basic idea is to look at the communication pattern generated by a particular host, and to compare it to the behavior patterns representing different activities/applications [12]. The connection patterns describe network flow characteristics corresponding to different applications by capturing the relationship between the use of source and destination ports, the relative cardinality of the sets of unique destination ports and IPs as well as the magnitude of these sets. The application specific behavior patterns are often difficult to find, especially if multiple application types are used simultaneously. In order to identify a communication pattern reliably, the method needs a lot of flows coming from and going to a host.

Authors of [12] validated their method by using signature based classification. As there are no commonly accepted and well performing byte-signatures, authors constructed their own signature database.

– **Statistics based classification:** In statistics based classification some statistical feature of the trace is captured and used to classify the network traffic. To automatically discover the features of a specific kind of traffic, the statistical methods are combined with methods coming from the field of artificial intelligence. The most frequently discussed method is the Bayesian analysis technique as in [14], [19], [13], [11], [10]. The basic requirement of these techniques is previously hand-classified network traffic which provides them with training and testing data-sets. In order to reach sufficient accuracy, the ratio of these data-sets should be about 1:1.

In [19] authors used port based classification to validate their method. They assume that for the ports they use in the study the majority of the traffic is from the expected application. In this case, it is most likely that few 'wrong' flows would decrease the homogeneity of the learned classes. Therefore their evaluation results can be treated as lower bound of the effectiveness. They also do not consider traffic of the selected applications on other than the standard server ports. Authors of [11] worked with commonly available traffic traces, but these traces contained only packet headers which excludes such reliable validation methods which are based on packet payload. In [10], the traffic classification method was applied online without capturing the original data due to the lack of capacity to store the massive amount of data which is the consequence of high traffic speeds. This makes impossible to validate the traffic classification by others.

– **Information theory based classification:** A useful aid in traffic classification is introduced in [18] which is an information theoretic approach and can group the hosts into typical behaviors e.g., servers, attackers. The main idea is to look at the variability or randomness of the set of values that appear in the five-tuple of the flow identifiers, which belong to a particular source or destination IP address, source or destination port. The information theoretic approach can not be used for flow level traffic classification in the same way as the other methods. It is just an aid in traffic classification and arises the problem that it can only specify very broad application types but not capable of classifying specific applications. This method intensively uses the five tuple identification of the flows without other additional information.

Authors of [18] validated the identified clusters by checking the found dedicated port of the hosts with the port-application database used for port based classification.

– **Combined classification method:** A couple of different approaches have been proposed in the literature for traffic classification, but none of them performs well for all different application traffic types present in the Internet. Thus, a combined method that includes the advantages of different approaches is proposed in [17], in order to provide a high level of classification completeness and accuracy. The classification method in [17] is based on a complex decision mechanism, in order to provide an appropriate identification mode for each different application type. As a consequence, the ratio of the unclassified traffic becomes significantly lower. Further, the reliability of the classification improves due to the joint decision of various methods.

Authors of [17] validated their method by comparing the results of the introduced method to the results gained from applying all the independent traffic classification mechanisms and their trivial combination on the same traffic traces. The used datasets are full packet length traces measured in several operational mobile broadband networks, but none of them publicly available.

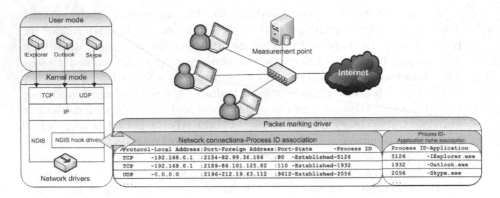

Fig. 1. The position of the proposed driver within the terminal

3   The Proposed Method for Validation

In this section we describe our proposed method for the validation of traffic classification algorithms. As we already mentioned before, instead of validating passive methods by each other we design an active validation method, because we look for a deterministic and reliable solution.

The principle of the method is the following: at the traffic generating terminal, packets are collected into flows and flows are marked with the identifier of the application that generated the packets of the flow. The two main requirements on the realization of the method are that it should not deteriorate the performance of the terminal, and the byte overhead of marking should also be negligible.

The preferred realization is a driver that can be easily installed on terminals. The position of the introduced driver can be seen in Figure 1. It takes place right before the network interface thus each packet exchanged between the terminal and the network has to pass through it. We have implemented a prototype, which is a Windows XP driver based on the Network Driver Interface Specification (NDIS) library. The kernel NDIS library abstracts the network hardware from network drivers and provides an API through which intelligent network drivers can be efficiently programmed. If the sending and receiving functions of the NDIS IP protocol driver are hooked, all TCP and UDP packets can be intercepted and filtered. This method lets developers create for example, firewalls, sniffers, traffic meters or network analyzers based on this technology.

To meet our requirements, the driver is designed to work in the following way. In the case of a passing through packet the following process takes place (see Figure 2):

1. The packet is examined whether it is an incoming or outgoing packet. In case of an incoming packet, the process ends without marking the packet as it is not beneficial to mark incoming packets.
2. In case of an outgoing packet, the size of the packet is examined. If the current packet size is already the size of Maximum Transmission Unit (MTU), the

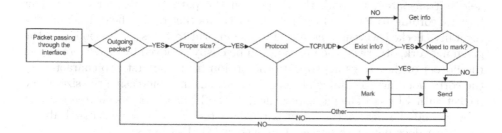

Fig. 2. The working mechanism of the introduced driver

extension of the packet with marking would lead to IP fragmentation. To avoid this, the process continues with only those packets which are smaller than the MTU decreased with the size of marking. Initiating messages in protocols are typically small e.g., the SYN packet of a TCP packet is only a flag, thus there is practically no loss (unmarked flow) with the introduction of this condition.

3. As there is no information in the operating system about those 'network connections' which use other protocols than TCP or UDP, the process continues with only TCP or UDP packets.

4. According to the five tuple identifier of the packet, it is checked whether there is already available information about which application the flow belongs to. The driver has to cache this information because querying the operation system about the existing network connections is very resource consuming and can not be done at high network speeds. We used a hash as the data structure for the cache as it can be directly addressed by the searched data. If there is no information on the flow in the cache yet, the operating system is queried to supply the established network connections and the process IDs of the responsible applications. The process IDs are state specific information in the operating system. To get a universal name about the application, the process IDs are connected to the application's executable name as can be seen in Figure 1.

5. When all information is prepared for the marking of the packet, there is a final chance to decide whether the driver should mark the packet or not. The packet marking can be done for all of the packets in the flow, randomly selected packets of the flow, only the first packet of the flow or it is also possible to switch off the marking for specific applications. There is an option for the random selection of packets to be marked to enforce the first packet of the flow to be marked or avoid the first packet to be marked. The sense of these options is to make an optimal trade-off between performance, network transparency and to ensure high chance of recordable marked packets in the case of network loss.

The marking is done by extending the original IP packet with one option field. We selected the Router Alert option field, because the existence of this field is transparent for both the routers on the path and also for the receiver host (according to RFC 2113 [3]). If one uses another option field, it should be carefully checked whether the marking is conform to the security policy of the given network, otherwise the marking can be easily removed by an edge router in the border of the access network. In the option field, the first two characters of the corresponding executable file name are added, thus increasing the size of the packet with 4 bytes. The packet size field in the IP header is also increased with 4 bytes and the header checksum is recalculated. As already discussed above, the driver does not mark packets larger than (MTU-4 bytes).

4   The Validation of a State-of-the-Art Traffic Classification Method

A reference measurement [7] has been created as a proof-of-concept of the introduced validation method. For the sake of simplicity, the measurement took place in a separated access network. Our driver has been installed onto all computers on this network. The duration of the measurement was 43 hours. The captured data volume in the network is 6 Gbytes, containing 12 million packets. The measurement contains the traffic of the most popular P2P protocols: BitTorrent, eDonkey, Gnutella, DirectConnect; VoIP and chat applications: Skype, MSN Live; FTP sessions, filetransfer with download manager; e-mail sending, receiving sessions; web based e-mail (e.g., Gmail); SSH sessions; SCP sessions; FPS, MMORPG gaming sessions; streaming radio; streaming video and web based streaming. In Figure 4 the traffic mix of the measurement can be seen. Both the volume and the flow number ratio of different applications is presented.

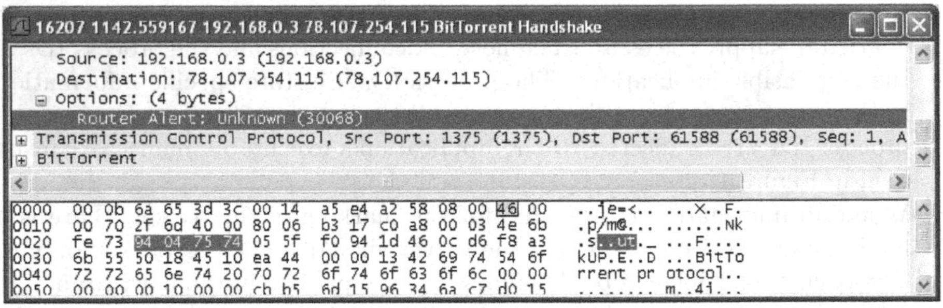

Fig. 3. A marked packet of the BitTorrent protocol

In Figure 3 an example of the marked packets can be seen. The IP header shows the increased size of the packet (without the option field, the value where currently is 46 would be 45) and the option field is highlighted, where the last two fields could be used to place the marking. The marking shows that the generating application was the uTorrent [8] BitTorrent client (by the first two characters in its name).

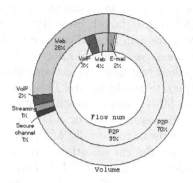

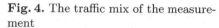

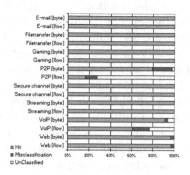

Fig. 4. The traffic mix of the measurement

Fig. 5. The results of the classification compared [17] to the reference measurement

The traffic classification method that we wish to validate is described in reference [17], with the addition that the classification of VoIP applications has been extended with ideas from [15] (see the discussion later below). In Figure 5, it can be seen that e-mail, filetransfer, streaming, secure channel, and gaming traffic has been identified very accurately. This is due to the fact that these applications use well-documented protocols, open standards, and do not constantly change. In the case of those protocols which use encryption, the session initiation phase is critical as this phase can be identified the most accurately. In such common protocols as SSH or SCP it can be done with full success, however in such proprietary protocols like Skype the identification fails for several flows.

In the case of classification of P2P applications there are several problems: one thing to note is that P2P applications created plethora of TCP flows containing 1-2 SYN packets probably to disconnected peers. This is the primary reason of the large number of unclassified P2P flows, while the unclassified P2P volume is low. As there is no payload in these packets, the signature based methods can not work. The flows are initiated from dynamically allocated source ports towards not well-known destination ports, thus the port based methods also fail. The server search and P2P communication heuristic [17] methods also fail because there are no other successful flows to such IPs.

Also some small non-P2P flows were misclassified into the P2P class. Fortunately, the number of such flows is small both in flow number and byte volume. We realized that the reason behind is the not fully proper content of the port-application database. Creating too many port-application associations easily results in the rise of the misclassification ratio.

The constant change of P2P protocols also causes some inaccuracy in the classification: there are new features added to P2P clients day-by-day, and their working mechanism can be typical for a selected client not the whole protocol itself.

Another problem of traffic classification is a matter of philosophy. There is traffic which is the derivation of other traffic: the simplest case is the DNS traffic which is the result of any traffic which uses domain names instead of specific IP addresses. For example, web creates DNS traffic though users do not want to create DNS traffic on purpose. There are more complicated cases: e.g., MSN uses HTTP protocol for transmitting chat messages, which do not need to be considered as web. Furthermore, the MSN client transmits advertisements over HTTP, but this cannot be recognized as deliberate web browsing. This raises the question whether such HTTP flows from the MSN application which are classified as web would have to be considered as misclassification, or it is acceptable that they are classified as web. In this comparison, to be fully objective, only that kind of traffic was considered as hit where the classification outcome and the generating application type (the validation outcome) agreed. For example, the chat on the DirectConnect hubs which has been classified as chat could have been considered as actually correct but in this comparison it was considered as misclassification.

The high VoIP hit ratio is due to the successful identification of both MSN Messenger and Skype. Skype is difficult to identify: for some of the Skype flows the problem is the same as in the case of P2P applications, further Skype is a proprietary protocol designed to ensure secure communication thus it is difficult to obtain a good protocol description. However, authors of [15] found a characteristic feature of Skype: the application sends packets even when there is no ongoing call with an exact 20 sec interval. In [17], there is a P2P identification heuristic which was designed to track any message which has a periodicity in packet sending thus the extension of the original method in [17] for the specific 20 sec periodicity of Skype was straightforward. The validation showed us the deficiency of the classification of Skype, thus with a simple extension of the algorithm it became proper for accurate Skype traffic identification as well. In this way the idea of [17] has been validated as it proved to be robust for the extension with new application recognition, and also the validation mechanism proved to be useful.

5 Summary and Future Work

In this paper we introduced a new active measurement method which can help in the validation of traffic classification methods. The introduced method is a network driver which can mark the outgoing packets from the clients with an application specific marking. With the introduced method we created a measurement and used this to validate the method presented in [17]. The method has been proved to be working accurately but also some deficiencies in the classification of P2P applications and Skype has been identified.

The introduced method can be used in several ways besides the main target of validating traffic classification. One straightforward continuation of this work is to use the marking method at the measurement side for online traffic classification (which we actually did during the debugging of the prototype). This assumes that the terminals accessing an operator's network are all installed with the proposed driver, and also that the driver is made tamper-proof to avoid users forging the marking. Such an online classification could be used for online clustering of the traffic into QoS classes based on the resource requirements of the generating application. It could also be used by operators to charge on the basis of the used application by the user. The marking could be extended by other information about the traffic generating application, e.g., version number, thus the operator could track the security risks of an old application.

Acknowledgements

We would like to thank the help of Péter Brezina in the development of the introduced driver and the support of his supervisor Sándor Molnár.

References

1. IANA.TCP and UDP port numbers,
   http://www.iana.org/assignments/port-numbers
2. MSN Messenger, http://join.msn.com/messenger/overview2000
3. RFC 2113, http://www.networksorcery.com/enp/rfc/rfc2113.txt
4. RFC 2246, http://www.ietf.org/rfc/rfc2246.txt
5. RFC 4251, http://www.ietf.org/rfc/rfc4251.txt
6. Skype, http://www.skype.com
7. The measurement created for this article,
   http://pics.etl.hu/~szabog/measurement.tar
8. uTorrent, http://www.utorrent.com
9. World of Warcraft, http://www.worldofwarcraft.com/index.xml
10. Bernaille, L., Teixeira, R., Akodkenou, I., Soule, A., Salamatian, K.: Traffic classification on the fly, vol. 36, pp. 23–26. ACM Press, New York, USA (2006)
11. Erman, J., Arlitt, M., Mahanti, A.: Traffic classification using clustering algorithms. In: Proc. MineNet 2006, New York, USA (2006)
12. Karagiannis, T., Papagiannaki, K., Faloutsos, M.: BLINC: Multilevel Traffic Classification in the Dark. In: Proc. ACM SIGCOMM, Philadelphia, Pennsylvania, USA (August 2005)
13. McGregor, A., Hall, M., Lorier, P., Brunskill, A.: Flow Clustering Using Machine Learning Techniques. In: Proc. PAM, Antibes Juan-les-Pins, France (April 2004)
14. Moore, A.W., Zuev, D.: Internet Traffic Classification Using Bayesian Analysis Techniques. In: Proc. SIGMETRICS, Banff, Alberta, Canada (June 2005)
15. Perenyi, M., Molnar, S.: Enhanced skype traffic identification. In: Proc. Valuetools 2007 (2007)
16. Sen, S., Wang, J.: Analyzing peer-to-peer traffic across large networks. In: Proc. Second Annual ACM Internet Measurement Workshop (November 2002)
17. Szabó, G., Szabó, I., Orincsay, D.: Accurate traffic classification. In: Proc. IEEE WOWMoM, Helsinki, Finnland (June 2007)
18. Xu, K., Zhang, Z., Bhattacharyya, S.: Profiling Internet Backbone Traffic: Behavior Models and Applications. In: Proc. ACM SIGCOMM, Philadelphia, Pennsylvania, USA (August 2005)
19. Zander, S., Nguyen, T., Armitage, G.: Automated Traffic Classification and Application Identification Using Machine Learning. In: Proc. IEEE LCN, Sydney, Australia (November 2005)

Evaluation of Header Field Entropy for Hash-Based Packet Selection

Christian Henke, Carsten Schmoll, and Tanja Zseby

Fraunhofer Institute Fokus, Berlin, Germany
{christian.henke,carsten.schmoll,tanja.zseby}@fokus.fraunhofer.de

Abstract. Network Measurements play an essential role in operating and developing today's Internet. High data rates and complex measurement demands can origin an immense resource consumption for measurement tasks. Data selection techniques, like sampling and filtering, provide efficient solutions for reducing resource consumption while still maintaining sufficient information about the metrics of interest. Hash-based packet selection allows a synchronized selection of packets at multiple observation points. With this, the tracking of the path of a packet and the calculation of multipoint QoS metrics like one-way delay becomes possible. Nevertheless, hash-based selection is deterministic based on parts of the packet content and hence it is suspect to bias. The packet content used for hashing is a source for bias if the selected content is not variable enough. This paper empirically analyzes which header bytes are most variable and recommendable as input for hash-based selection if one targets the emulation of random selection.

1 Introduction

There exists a variety of applications for multipoint network measurements. Service Providers need to validate their delay guarantees from Service Level Agreements and network engineers have incentives to track where packets are changed, reordered, lost or delayed, for instance in error-prone environments like Mobile-Adhoc Networks.

Hash-based selection is a passive measurement technique that enables multipoint measurements [1] [2] and packet tracing [3] [4]. In contrast to active multipoint measurements [5] [6] hash-based selection can calculate one-way metrics like delay without introducing additional traffic into the network. Hash-based selection reduces calculation effort compared to other passive multipoint measurements [7] [8] because it does not correlate all packets from the measurement points. Instead, hash-based selection only samples a consistent subset of packets at every measurement point and estimates the real traffic characteristics.

Hash-based selection is realized by the following technique. Parts of the packet content that are invariant between measurement nodes are extracted and used as the hash input for a hash function. The hash function with a digest length of N bits maps the hash input to a value in the hash range $R = [0..2^N - 1]$. The packet itself is selected if the hash value falls into a predefined selection range

M. Claypool and S. Uhlig (Eds.): PAM 2008, LNCS 4979, pp. 82–91, 2008.
© Springer-Verlag Berlin Heidelberg 2008

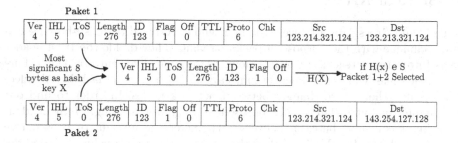

Fig. 1. Input Collision of 2 Different Packets

$S \subset R$. In order to obtain a certain sample size one can adjust the selection range. The advantage of this technique is that the selection decision for each packet along its path is the same, assuming that the selected packet content (hash input), hash function and selection range are the same at the different measurement points.

2   Problem Statement

Hash-based selection is solely based on the packet content. Therefore it is suspect to bias, i.e that the distribution of attributes in the sampled subset is different to the one in the population. E.g. if packets with certain length are preferred in the selection, a sound estimation of the packet length distribution and related properties is impossible. The choice of the packet content [1] to be used for the hash input is of utter importance because this decision can introduce bias. This is shown in the example Fig. 1 which shows two similar packets. We assume that only the first 8 bytes of the IP header are used as the hash input. Because these fields are equal, both packets form the same hash input X. Thus both have the same hash value H(X) and same selection decision. Assuming there are more packets like Packet 1 and 2 that are equal in the most significant 8 bytes, they will all map to the same hash value and either all or none of them are selected. This means that packets with 276 bytes length are either over- or underrepresented in the sampled subset. The selected bytes from the packet content should be sufficient to identify unique packets in order to avoid *input collision*, i.e packets with the same hash input. It is not suitable to use the whole packet content as hash input because the calculation time for hash functions increases significantly with increasing hash input length as shown in [9]. The hash function has to be applied for each packet which would deplete the processing resources of the measurement node in case long hash inputs are used. In this paper we empirically evaluate the IPv4, IPv6 and transport header fields on their suitability as hash input in hash-based selection and give recommendations on which ones to use.

---

[1] IN PSAMP terms the packet content consists of the IP header and payload.

3   State of Art

Duffield and Grossglauser [3] were the first who introduced the hash-based selection technique with the purpose of packet tracing. They divide the IPv4 header fields into three categories, variable, low entropy and high entropy fields. Further, fields like TTL and IP checksum can not be used for hash-based selection because their value changes between network nodes. In order to find out the amount of bytes that should be used as input for hashing, Duffield identifies input collisions between packets with increasing amount of consecutive bytes from the packet. Molina [10] evaluates how many hash values collided depending on the input length. Snoeren [4] uses the hash-based approach for IP traceback systems and identifies unique packets by considering different input lengths. All three authors conclude that an input length between 20 and 40 consecutive bytes is sufficient for hash-based purposes. Gong Jian [11] analyzes the entropy of IPv4 fields per bit and byte on a CERNET backbone link in order to find the most variable fields encouraging the idea of packet selection based on the value of the IP Identification field. Nevertheless in [2] Zseby reasons that the IP ID field alone is not sufficient to identify unique packets due to inconsistent IP ID field handling of different operating systems. The Packet Sampling Working Group (PSAMP) [12] recommends the use of these header fields: ID, Flags, Fragment offset, source and destination address and a configurable number of bytes from the IP payload, starting at a configurable offset. Duffield, Snoeren and Molina only consider general assumptions about header fields that are suitable for hash-based selection but lack the basis of an empirical and bytewise analysis. Our approach based on entropy measurements enables us to rate header bytes according to their suitability for hash-based selection. We use this systematic approach in order to find a smaller subset of bytes that show similar or better results than the recommended PSAMP bytes.

4   Approach

4.1   Header Fields Properties

The header fields need to be *1) static between network nodes* because the selection decision based on this content is required to be consistent throughout the network. Header fields that are not static between network nodes are: Time to Live and IP Checksum. On the other hand, the header fields need to be *2) variable among packets* in order to prevent bias caused by input collisions, i.e. the hash input of every unique packet should be unique. The higher the variability in one header field the higher the probability that two packets differ in this field. In order to assess each packet bytes' variability, we calculate the entropy of each byte.

Table 1. Traces Used for Evaluation

Trace Name	Location	Duration	IP Snapsize	packets in millions	IP Address anonymous	IP Version
NZIX	New Zealand	30 hours	40	~200	Yes	4
Ciril	France	6 hours	20	~100	No	4
FH Salzburg	Austria	3 days	40	~110	Yes	4
LEO 1		3 hours	140	~130	No	4
LEO 2		6 min	140	~12	No	4
Twente	Netherlands	10 days	40	~380	Yes	4
Mawi	WIDE-6Bone	40 days	60	~80	Yes	6

4.2 Entropy

The entropy per byte is defined as [13]:

$$H(B) = -\sum_{i=0}^{255} p_i log p_i \qquad (1)$$

Where B is the discrete variant of byte values. p_i is the probability that the byte value i occurs. For comparison purposes we divide the byte entropy by its maximum value H_{max} to obtain the information efficiency E [11].

$$E = \frac{H(B)}{H_{max}(B)} \qquad H_{max}(B) = -\sum_{i=0}^{255} \frac{1}{256} log_2 \frac{1}{256} = 8 \qquad (2)$$

The information efficiency E measures the byte randomness. An information efficiency of 0 denotes no randomness (i.e only one value occurs) and a value of 1 denotes maximum entropy with equally distributed byte values.

4.3 Traces Used

We evaluated a set of 7 trace groups (cf. Tab. 1). All traces can be accessed via the MOME database [14]. The NZIX traces were captured by the WAND research group at a a peering point among a number of major ISP's. FH Salzburg traces were captured at an WAN Access Network on a student campus. The Twente traces were captured at an aggregated uplink of an ADSL access network. There are two trace groups that were captured by a large european telecom operator and are noted as LEO1 and LEO2. The Mawi traces were the only

Table 2. Traces Protocol Mix in %

	TCP	UDP	ICMP	others	Main Applications
FHSalzburg	99	1	0	0	http(90)
LEO 1	90	10	0	0	edonkey(25) http(5)
LEO 2	33	60	0	7	tunnel(60) edonkey(10)
NZIX	68	20	9	3	http(50) quake(5)
Twente	89	7	1	3	diversified

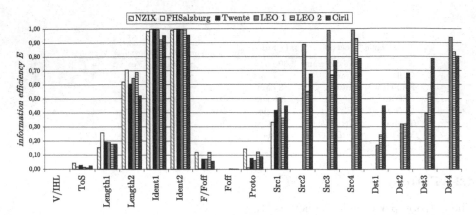

Fig. 2. Information Efficency IPv4

accessible IPv6 traces. All but the Ciril and IPv6 Mawi traces include parts of the transport header. Table 2 shows that each trace group represents a different scenario indicated by their different transport protocol mix and applications. We assessed the applications by the transport header port addresses.

5   Entropy Evaluation

The entropy per byte is calculated with a tool written in C++ which uses the libpcap library to read packet traces. The tool calculates the frequencies of each byte value and calculates the entropy.

5.1   IPv4

The IP source and destination addresses from Twente, NZIX, FHSalzburg traces are anonymized and do not reveal the real entropy and hence are not evaluated. Only the Twente and FHSalzburg traces were anonymized with the tcpdpriv "-A 50" option which preserves the entropy in the most significant address byte. Figure 2 shows the results but does not include the Time To Live and Checksum field as these are variable from hop-to-hop and can not be used for hash-based selection. Version, IP Header Length and the least significant byte (LSB) of Fragment Offset are not shown as well, as their entropy is very close to zero. For multiple byte header fields the LSB is depicted on the right. The identification (ID) field has the highest entropy. One could assume that the source and destination address and the ID is sufficient to distinguish packets, but the ID field bears a problem: we observed many packets with an ID of 0 which resulted from different IP ID handling of operating systems.

Since there exist many small packets shorter than 255 bytes, the entropy for the most significant byte (MSB) of the length field is comparably low (about 0.2 information efficiency). The figure shows that there is an increase of entropy from MSB to LSB of the IP source and destination addresses. Nevertheless, it is expected that the information efficiency of all 8 bytes is significantly less than all its components, because there is a strong dependency between the address bytes. The highest average information efficiencies can be found in these bytes: *ID, Source and Destination Address (two LSB), Length(LSB).*

5.2 TCP

For analyzing the transport header fields on their entropy, the Ciril and Mawi traces had to be omitted, because they do not include any transport header data. All header fields are applicable for hash-based selection because they are invariant between network nodes and their entropy results are shown in Fig. 3. The most suitable fields for hash-based selection with an information efficiency close to 1 are: *Checksum, Sequence Number and Acknowledgment Number.* The FHSalzburg traces show comparably low entropy in the acknowledgment number field as many packets have an ack of zero.

5.3 UDP

The entropy results for the 8 byte UDP header are shown in Fig. 4. In the LEO2 trace group the checksum is set to zero for most packets maybe due to security reasons. Most of the packets with no UDP checksum include the layer 2 tunnel protocol with IP and TCP in its payload. The packets including the tunnel protocol use the same source and destination port numbers 1701 which causes the low entropy for the LEO 2 traces in port addresses as well. The layer 2 tunnel protocol is used in virtual private networks. Although the checksum is optional it is still advisable to use the UDP *Checksum field* for hash-based selection as it includes very high entropy for all but the LEO2 traces. Other fields with high entropy are: *SrcPort (LSB), DstPort (LSB), Length (LSB).*

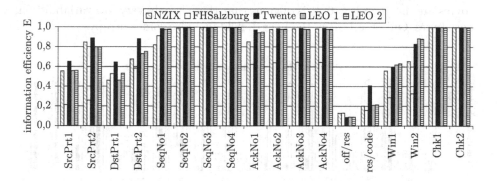

Fig. 3. Information Efficiency TCP

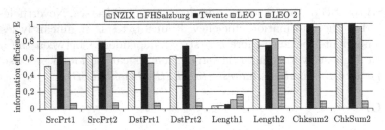

Fig. 4. Information Efficiency UDP

5.4   ICMP

The Internet Control Message Protocol (ICMP) differs in length and content dependent on the meassage type. All ICMP packets have four bytes in common: 1 byte type, 1 byte code, 2 bytes checksum. Because there are only few ICMP packets in the analyzed traces we refrained from distinguishing each ICMP message type to analyze the entropy. Figure 5 shows the results. Because the FHSalzburg traces did not include a sufficient amount of ICMP packets the entropy results per trace are very variable and are not shown in the diagram. The bytes with the highest information efficiency are: *Checksum, Bytes 12-13, Bytes 18-19*.

5.5   IPv6

The Mawi traces were captured at 2 different measurement points C(6Bone) and D(WIDE_6Bone) and we evaluated the entropy of the IPv6 header for each sampling point separately. We only consider the first header (and not optionally concatenated next headers). The Hop Limit field is similar to the Time To Live field of IPv4 and is decremented each network hop, therefore it can not be used for hash-based selection. Since the source and destination addresses are anonymized we can not measure the entropy of the original fields. The results for the remaining 7 bytes are shown in Fig. 6. It is obvious that there is only very low variability in those bytes. Only the LSB of the Length field includes some entropy.

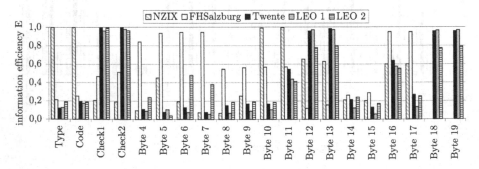

Fig. 5. Information Efficiency ICMP

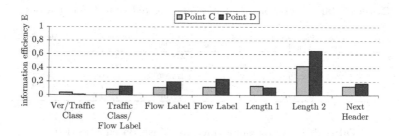

Fig. 6. Information Efficiency IPv6

6   Input Collisions

An input collision consists of multiple packets that have the same hash input.
As pointed out in Sect. 2, packets with same hash input and selection decision
introduce bias to the hash-based selection because these are either over- or underrepresented in the sampled subset. There are two reasons why packets hash
inputs' collide: 1) the packets are identical or 2) the packets are not identical
but the selected bytes for hash input. It is assumed that packets with identical headers are likely to be identical, because the transport header checksum is
calculated over the whole packet. For the presented evaluation all packets with
equal headers (IP and transport) are labeled to be identical. This is done despite the awareness that these identical labeled packets may not be identical (e.g.
for UDP packets that do not use the optional UDP checksum). The amount of
identical packets will be compared to collisions caused by a bad hash input configuration (only IP header without checksum and TTL), an 8 high entropy bytes
hash input configuration and a 16 bytes combination used by Molina in [10].

1. Recommended 8 Bytes - based on our entropy evaluation results we chose the
   IP ID field and 6 Bytes depending on the transport protocol: TCP (Checksum, 2 LSB of Sequence and Acknowledgment Number) UDP (Checksum,
   Source Port, LSB Destination Port, LSB Length) ICMP (Checksum, Bytes
   12,13,18,19)
2. Molina's 16 bytes - Molina [10] proposes the use of 16 bytes: Length, ID,
   Source and Destination IP address and the 2nd 32 bit word of the transport
   header.

6.1   Comparison of Header Byte Combinations

The size of a collison is the amount of packets with the same hash input. Large
collisions are of more concern than small collisions, because the colliding packets
in large collisions all have the same selection decision and are either over- or
underrepresented in the selected subset. The same amount of packets within
several small collisions are less crucial, because the packets from different small
collsions can have different selection decisions and there is no "all or none"
decision. Hence we only look at the 20 largest input collisions in each trace (cf.

Table 3. Input Collisions for Different Selected Byte Combinations

Trace Group	Trace Files	Packets/File in millions	Identical IP + Transport header	Identical IP Header	Recommended 8 Bytes	Molina's 16 Bytes
FH Salzburg	18	6	3,547	238,174	3,547	3,547
NZIX	19	10	484,034	1,564,246	484,405	1,562,066
Twente	36	10	13,120	475,570	16,004	49,477
LEO 1	12	10	61,072	450,273	73,730	86,809
LEO 2	1	10	949	8,116	7,919	1,121

Table 4. Bytes Recommended for Hash-Based Selection

IPv4	IPV6	TCP	UDP	ICMP
Identification	Length LSB	Checksum	Checksum	Checksum
Destin. 2LSB		Sequence No	Length LSB	Byte 12-13
Source 2LSB		Acknowledg. No	Source Port LSB	Byte 18-19
Length LSB			Destin. Port LSB	

LSB - Least Significant Byte

Tab. 3). Because the Mawi and Ciril do not include transport headers we could not evaluate them. For all traces we observed that there is a great amount of packets that occur in bursts and have different TTL and IP Checksum values but are equal in the remaining bytes. These packets are caused by routing loops or the lack of the IP ID field (IP ID=0). The Twente trace group consists of 36 trace files with 10 million packets each. The 20 largest collisions of each trace include 13,120 packets because the packets are identical. Using only the IP header as hash input we observed about 475,000 packets in the largest collisions for the Twente traces. With the use of our recommended 8 Bytes 16,004 packets collided, whereas with Molina's 16 bytes 49,477 packets did collide. For the FHSalzburg trace group, the content combinations of Molina and ours identify the exact amount of identical packets. Our subset shows a similar amount of identical hash inputs for the NZIX traces whereas Molinas' subset has 3 times more collisions. The LEO2 trace group that consists of VPN tunnel traffic shows a different result: Molina's 16 byte subset includes less collisions. This is caused by the low entropy of our recommended bytes of the LEO2 UDP header fields (see Sect. 5.3) especially the non-utilized UDP checksum.

7   Conclusion

From our analysis we conclude that there are certain bytes in the IP and transport layer header that are more suitable for hash-based selection than others because they are more variable and help to distinguish unique packets. Although there are some high variable fields in the IPv4 header they are not sufficient for the hash input in hash-based selection. Hence one has to include more variability by using transport header fields. The TCP header includes many high entropy bytes, whereas UDP and ICMP lacks some. The ICMP packets are usually less observed in a packet stream but because of the similarity and the short length ICMP packets easily collide in the hash input if not enough bytes are used. As

ICMP and UDP packets are rare but strongly similar one may even consider to use an additional amount of bytes for the different protocols in order to enable a better differentiation of packets.

We evaluated IPv6 header fields on their suitability for hash-based selection. The IP address fields could not be evaluated because of anonymization. Nevertheless it is reasonable that it is insufficient to use only the 40 byte IPv6 Header because at one observation point many packets will include the same destination and source address. Other bytes do not show high variability and cannot help to distinguish packets. In future it has to be evaluated if the transport headers used with IPv6 do behave the same as IPv4 transport headers.

We observed in our input collision evaluation that the selection of high entropy bytes as hash input can decrease the amount of bytes required for hash-based selection. Although we only used 8 bytes compared to Molina (16 bytes) we gained better results for all trace groups except for LEO2. This more than halves the hash calculation time [9] for the BOB hash function (recommended by PSAMP).The LEO2 consists of UDP VPN traffic with many packets without UDP checksum and low variable UDP header fields that causes additional collisions with our recommended 8 bytes. Our recommended combination is sufficient for most traces but can produce additional input collisions. In order to improve the results one can add more high entropy bytes from those proposed in Table 4.

References

1. Niccolini, S., Molina, et al.: Design and implementation of a one way delay passive measurement system. In: Network Operations and Management Symposium (2004)
2. Zseby, T., Zander, S., Carle, G.: Evaluation of building blocks for passive one-way-delay measurements. In: PAM Workshop, Amsterdam, Netherlands (April 2001)
3. Duffield, N., Grossglauser, M.: Trajectory sampling for direct traffic observation. IEEE/ACM Trans. Netw. 9(3), 280–292 (2001)
4. Snoeren, A., Partridge, C., et al.: Single-packet ip traceback. IEEE/ACM Trans. Netw. 10(6), 721–734 (2002)
5. Active measurement project, http://amp.nlanr.net/
6. CAIDA. Skitter, http://www.caida.org/tools/measurments/skitter/
7. Papagiannaki, K., Moon, S., et al.: Analysis of measured single-hop delay from an operational back bone network. IEEE Infocom, New York (June 2002)
8. Choi, B.Y., Moon, S., et al.: Practical delay monitoring for ISPs. In: ACM Conference on Emerging network experiment and technology. ACM Press, New York (2005)
9. Henke, C., Schmoll, C., Zseby, T.: Empirical evaluation of hash functions for multipoint measurements. Technical Report TR-2007-11-01 (Available upon request)
10. Molina, M., Niccolini, S., Duffield, N.G.: Comparative experimental study of hash functions applied to packet sampling. In: ITC-19 (August 2005)
11. Jian, G., Guang, C.: Distributed sampling measurement model in a large-scale high-speed ip networks. Journal of Southeast University, Nanjing, China (2002)
12. Zseby, T., Molina, M., et al.: Sampling and filtering techniques for IP packet selection. In: IETF Internet Draft (2007)
13. Bronstein, I.N.: Taschenbuch der Mathematik Teubner, Leipzig (1962)
14. Traffic measurement database MOME, http://www.ist-mome.org/

A Reactive Measurement Framework

Mark Allman and Vern Paxson

International Computer Science Institute

Abstract. Often when assessing complex network behavior a single measurement is not enough to gain a solid understanding of the root causes of the behavior. In this initial paper we argue for thinking about "measurement" as a *process* rather than an *event*. We introduce *reactive measurement* (REM), which is a technique in which one measurement's results are used to automatically decide what (if any) additional measurements are required to further understand some observed phenomenon. While reactive measurement has been used on occasion in measurement studies, what has been lacking is *(i)* an examination of its general power, and *(ii)* a generic framework for facilitating fluid use of this approach. We discuss REM's power and sketch an architecture for a system that provides general REM functionality to network researchers. We argue that by enabling the coupling of disparate measurement tools, REM holds great promise for assisting researchers and operators in determining the root causes of network problems and enabling measurement targeted for specific conditions.

1 Introduction

Because networks are vast collections of integrated components, it can often be the case that analyzing some network behavior in depth (for characterization, tuning, or troubleshooting) requires adapting on-the-fly what sort of measurements we conduct in consideration of the conditions manifested by the network. While the technique of adapting measurements dynamically has been recognized by practitioners in a number of contexts, a key missing element has been the ability to tie together disparate forms of measurement into a cohesive system that can automatically orchestrate the use of different techniques and tools.

To this end, we outline a new measurement paradigm: *reactive measurement* (REM). The vision of REM is to provide a platform that can *couple* measurements—both active and passive—together in a way that brings more information to bear on the task of determining the root cause of some observed behavior. For instance, consider the problem of analyzing the failure of a web page to load. When a REM system observes unsuccessful web page requests, it can automatically execute a set of diagnostic measurements designed to winnow the set of possible reasons for the failure down to the root cause(s) (e.g., a subsequent *traceroute* may highlight a disconnect or loop in the path). While any particular reactive measurement task can be manually pieced together with straightforward scripting, many of the tasks (collecting events, expressing dependencies, managing timers, archiving results to varying degrees) benefit a great deal from a "toolbox" approach. Essentially, it is the absence of such a toolbox that, we believe, has led to a failure to exploit reactive measurement to date.

M. Claypool and S. Uhlig (Eds.): PAM 2008, LNCS 4979, pp. 92–101, 2008.
© Springer-Verlag Berlin Heidelberg 2008

The basic notion behind the reactive measurement paradigm is that automatically coupling disparate measurement techniques can bring more information to bear on the task of gaining insight into particular network behavior. The fundamental REM building block is having one measurement's result trigger additional *reactive measurements*. Thus, when a particular behavior is observed, we can automatically trigger additional measurements to work towards determining the root cause(s) of the behavior. Furthermore, as those tools hone in on the underlying reasons—or determine that a given hypothesis is incorrect—their output can again trigger the additional measurements needed to drive progress forward. The paradigm of reactive measurement is to think of "measurement" as a process rather than a simple activity. The goal of the process is to gain insight, and in a system as complex as today's networks such a task will likely involve more than one assessment technique.

This quite simple idea holds promise both for providing a foundation for significant advances in network troubleshooting, and for fostering new types of Internet measurement studies. Regarding this latter, the literature is filled with Internet measurement studies that evaluate the behavior of networks that are working as expected. These studies sometimes offer glimpses of the failure modes present in the current network, when such glitches are observed in the course of taking measurements (e.g., [12] identifies routing "pathologies", which are then removed from subsequent analysis). REM, however, enables the opposite approach. Because REM can key on *anomalies* in the network, REM can be used to trigger measurement infrastructure precisely when unexpected events occur, enabling us to learn a wealth of information about the causes of the problems and their immediate effects. We can further ultimately envision REM as the basis for networks that can automatically diagnose problems and take steps to work around detected failures.

REM enables fundamentally new ways to measure network behavior that cannot be accomplished with stand-alone active or passive techniques. Consider the case of measuring failures in the Domain Name System (DNS). While a number of studies on the operation of the DNS have been conducted (e.g., [6]), the fundamental question "how long does a particular DNS failure persist?" remains largely unanswered. This question cannot be answered by simply monitoring the network, because the experiment is then beholden to users who may or may not trigger additional DNS requests after a failure (particularly if they've been trained by the failure patterns they've experienced in the past). Alternatively, researchers could actively query the DNS for a set of hostnames independent of the requests invoked by actual users. In this case, following up on failed requests is straightforward. However, while this approach can shed light on the original question, the workload imposed on the DNS and network is synthetic and likely unrealistic. Using reactive measurement allows for bringing both active and passive measurements to bear to answer the basic question: a monitor can observe naturally occurring DNS requests in the network, and, upon noticing a failed DNS request, the REM system triggers an active measurement tool to periodically query the DNS to determine how long the failure persists, whether the failure is intermittent, etc. We can also invoke additional tools to determine *why* the DNS requests are not completing.

A second use of REM is for *targeting* measurements. Consider a packet-trace study investigating the behavior of networks and protocols under "very congested" conditions to gain insight into how to evolve protocols and algorithms to work better in such situations. The way this is often done today is to trace the network for a lengthy block of time and then post-process the resulting traces for periods when the network is "very congested," discarding the remainder of the trace. This methodology is scientifically sound, but logistically cumbersome due to the volume of traces that must be initially collected. Using a REM system, however, the researcher could first passively assess the state of the network, and then trigger detailed packet capture only when the network is in the desired state. In this way, not only does the researcher not have to capture and store traffic that will ultimately not be used, but the traffic that is captured is immediately available for analysis without pre-processing. In this case, REM does not provide a methodology for conducting a fundamentally different experiment than could otherwise be undertaken (as is the case for the DNS investigation described above), but it eases some key logistical challenges by providing targeted measurements. *This is not a minor benefit, as the logistical burdens can easily be such that they, in fact, provide the ultimate limit on how much useful data is gathered.*

Finally, we note that while we have framed the REM system in terms of reactive *measurements*, the system is general enough to support a much broader notion of a *reaction*—such as something that is executed, but is not a measurement. For example, a generic reaction could page a network operator when the system has determined that a router has crashed. Ultimately, the REM system could be used as a platform to automatically *mitigate* or *correct* observed problems. For instance, if the REM system determines that a local DNS server has crashed, it could trigger a backup server to take over (as well as notifying operators of the change). Using the REM framework in this way offers great potential for providing a powerful method to add robustness to networks.

The remainder of this paper is structured as follows. We sketch related work in § 2. In § 3 we present the architecture of a prototype REM system that we have developed to support diverse measurement needs by providing the "glue" with which to tie together arbitrary active and passive network measurement tools. We briefly summarize in § 4.

2   Related Work

First, we note that the wealth of work the community has put into developing active and passive measurement tools forms a *necessary* component of the REM framework. As outlined in this paper, the reactive measurement system conducts no measurements itself. Rather, it leverages the results from independent active measurement tools and passive traffic monitors as input into a decision process as to what subsequent measurements are required to uncover the cause(s) of a given network phenomenon.

Many past studies have employed multiple measurement techniques in an attempt to gain broader insight on a particular problem than can be obtained when using a single measurement method. For instance, [9] uses both *traceroute* and BGP routing table analysis to determine the AS path between two given hosts. The key difference between these kinds of studies and the REM framework outlined in this paper is in

REM's *automated coupling* of measurements. REM specifically defines dependencies between the output of a measurement tool and what (if any) additional measurements are required. We note that REM is orthogonal to and does not obviate the usefulness of studies like [9] that leverage information from multiple independent measurements.

The literature also has examples of researchers utilizing the reactive measurement notion. For instance, [3] uses *traceroute* measurements to followup on the detection of possible "missing routes" found by analyzing BGP routing tables. Another example is discussed in [2], whereby incoming email is first classified as spam or ham and then the URLs within the spam are followed in an effort to characterize various scams. While researchers have used REM techniques in the past for specific purposes, what has been missing is to systematize these mechanisms in order to make REM broadly available to the research community as a general approach.

In addition, we note that our framing of measurement as a process rather than an event shares some properties with PDA [5] (which is mainly focused on host problems, but does touch on connectivity issues as well), ATMEN [8] (which is largely concerned with coordinating distributed triggered measurements across organizations), and the general idea of "trap directed polling" via SNMP information. All of these systems in some fashion make use of one measurement to drive another measurement (and/or ultimately make a conclusion), but all focus on different aspects of the problem.

Reactive measurement shares some of the goals of the "knowledge plane" (KP) proposed in [4]. The KP envisions continuously gathering information about the network. When particular behaviors need further investigation the KP can be queried to gain a breadth of relevant information. One immediate and practical problem with the KP approach is the immense task in gathering and sifting through information about the entire network. REM proposes essentially the opposite approach: rather than synthesizing from already-gathered information, REM aims to *adaptively* gain insight into particular observed behaviors by running a series of measurements in response to a given phenomena. REM thus has the advantage that it can be conducted *locally*. No distributed data substrate—with the attendant difficulties of scaling, privacy, security, trustworthiness—needs to be constructed. That said, we note that REM in some sense is also orthogonal to ambitious approaches such as KP. The two could be coupled, such that facts learned by REM activity are fed into the KP data substrate, and REM itself could incorporate facts extracted from the substrate to drive its local decision process (as discussed in more detail in § 3.4).

Finally, we note that intrusion detection systems (IDS) share some high-level notions with REM [18,13]. IDS systems passively observe traffic to draw observations regarding network activity. These observations can be hooked to a "reaction", ranging from logging an event to resetting a TCP connection to adding a firewall rule to block traffic from a host that is port scanning the network. The REM concept of a reaction is much broader than the security-related reactions that popular IDS systems incorporate. In addition, IDS systems offer a passive view of the network, while reactive measurement allows for active probing to determine the state of the network. However, the ability of some IDS's to sift through large traffic streams to find specific types of high-level activity offers great promise of leverage within the REM framework (see § 3).

3   REM Architecture

This section presents an architecture for a generic, reusable reactive measurement system suitable for a broad array of measurement efforts. Our aim is to both explicate the approach and solicit input from the community while the effort is in its formative stages. We begin with a discussion of incorporating external measurement tools into the system. We then present the internal machinery that drives the measurement procedures, briefly delving into some of the details. Finally, we discuss possibilities for integrating the REM system with other external resources.

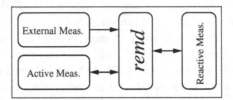

Fig. 1. Conceptual layout of the reactive measurement system

3.1   External Interactions

Fundamentally, the REM system couples measurements with reactions. Figure 1 illustrates the system's basic structure: arbitrary measurement tools glued together using a daemon, *remd*, that can be run on any general purpose computer connected to the network to be measured. *remd* provides an interface to and from traditional measurement tools, as well as a method for specifying the relationships between the measurements (outlined in the next subsection). First, we outline the various measurements shown in Figure 1 with which *remd* interacts:

- **Active Measurements.** *remd* can initiate independent active measurements based on a run-time-configured schedule and incorporate their results as input into whether or not to follow up with a reactive measurement, and in what form. For instance, a simple *ping* measurement may be executed every N seconds, with various pieces of information returned (e.g., success/failure, loss rate, presence of reordering, etc.) to *remd*. These results could then trigger additional measurements in an attempt to determine the reason behind the initial observations.
- **Reactive Measurements.** These are measurements that *remd* executes in response to previously-measured network phenomenon. For instance, if a tool reports to *remd* that the loss rate between the local host and a given remote host exceeds a threshold, then a reactive measurement can be triggered to attempt to determine the cause of the increased loss rate or how long it persists. The results of reactive measurements are fed back to *remd* and are then used to determine whether further reactive measurements are needed. Reactive measurements can be active or passive measurements.

– **External Measurements.** These are measurement results delivered to *remd* without *remd* initiating them itself. These measurements could come from SNMP monitoring systems, routers, intrusion detection systems (IDS), system log analyzers or custom built monitors. Each of these entities potentially has a unique and useful vantage point from which to assess certain network conditions and attributes.

The various components of the system interact by passing structured messages between the *remd* and the measurement tools. We can incorporate arbitrary tools into the system by writing simple wrapper scripts[1] that (i) understand and process requests formed by the *remd*, (ii) evaluate the output of the given tool(s) (return codes, output files, standard output), and (iii) form responses in the format *remd* requires. We use XML for requests and responses to ensure an extensible message structure that can accommodate communication with arbitrary measurement tools (their diverse set of arguments and result types). In addition, XML parsers are widely available allowing users to construct wrapper scripts without building complicated parsers and in a wide variety of languages. Finally, we note that while the contents of the messages passed between the *remd* and the various measurement tools must be well-formed, the meaning of the information and its relationship within the overall experiment is defined at run-time by the *remd* configuration, allowing a great degree of flexibility and leaving *remd* as neutral glue.

3.2 Internal Architecture

Internally, the REM system has three basic components: a measurement scheduler, an event receptor, and a state machine to capture the linkages between measurements. The measurement scheduler runs measurement tools at prescribed times. For instance, the user may want to run a simple measurement to assess a path periodically, along with successive reactive measurements as dictated by the results of the first measurement. Or, upon detecting a failure the user may wish to run the reactive measurements after a given amount of time, rather than immediately (e.g., to test DNS resolution N seconds after observing a failed lookup). The event receptor receives notifications from external monitors (e.g., an SNMP monitor) that then may initiate a chain of reactive measurements, and from the activity of the reactive measurements themselves. Finally, the state machine manages the transitions between various measurements.

Figure 2 gives an example of an REM state machine. It codifies that REM should start a *ping* measurement based on an internal timer. Based on the results of the *ping* measurement, REM will execute zero, one, or two reactive measurements. If the loss rate measured by *ping* exceeds a threshold T, REM executes *treno* [10] in an attempt to determine where in the path the congestion occurs. If the *ping* measurement observes packet reordering on the path, REM uses *cap* [1] to assess the impact of reordering on TCP's congestion control algorithms. Note: if the *ping* indicates a loss rate that exceeds T and packet reordering is present both *treno* and *cap* will be executed (bringing up a number of coordination issues that we discuss in more detail below). In the case where the *ping* measurement indicates both a loss rate below T and no reordering, then no

---

[1] The NIMI measurement infrastructure [15,14] has successfully used a similar wrapper script technique to incorporate arbitrary measurement tools.

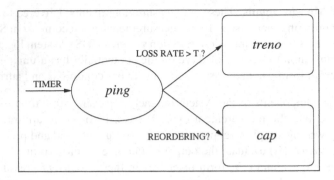

Fig. 2. Simple state machine whereby all measurements are invoked by *remd*

further measurements are executed. In other words, an implicit terminal state follows each state in the machine. If, after executing a measurement, none of the transitions are valid, then the current measurement chain ends. Finally, the *treno* and *cap* states could, of course, also have transitions to additional measurements.

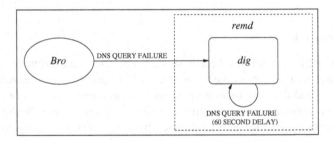

Fig. 3. Simple state machine of a reactive measurement triggered by an external monitor

Figure 3 gives a second example of an REM state machine. Here, *remd* (everything within the dotted line) receives a DNS failure notification from an external source, namely an instance of the *Bro* IDS (which can perform extensive, application-layer analysis of traffic). Upon receiving the message indicating a DNS failure occurred, *remd* executes a *dig* measurement in an attempt to resolve the given hostname. Each time the given hostname cannot be resolved, REM schedules another *dig* measurement for 60 seconds into the future. In addition to setting a time between measurements, a maximum number of attempts can be configured. For instance, inserting 60 seconds between DNS queries and running a maximum of 10 queries may suffice for a given experiment. Of course, a simple periodic timer will not suffice for all situations; our prototype REM system also provides Poisson-based intervals and exponential backoff.

The above examples are clearly simplistic, and the thorny problem of measurement scheduling and collision remains. A user may wish to have two reactive measurements run in parallel in one instance, and serially in another. In addition, a user may wish to base a reaction on the output of multiple measurements. These situations greatly complicate state machine construction. While this complexity can be hidden from the user by providing a high-level interface from which the system then creates the actual state machine, we may need a more powerful abstraction to cover all possible cases in the future. For example, we could use the quite general framework of Petri Nets [16] to codify the reaction path. Alternatively, we could directly employ Bro's events and timers. Our current prototype is based on a simple state machine. As we explore the sorts of reactive measurements we find we want to express in practice we will look to enhancing the system's abstract model to support these sorts of richer couplings.

As indicated above, external notifications to the REM system can come from any network monitoring system (IDS, SNMP, custom developed, etc.). Attempting to interface *remd* with legacy systems may require a lightweight shim to provide the necessary "plumbing". For example, consider integrating the *Bro* IDS into the REM framework. Since the *Bro* system includes a client library for transmitting *Bro* events and typed values, to integrate it with REM we can devise a simple event receiver that understands *Bro* events and translates them into *remd* notifications.

Note, as discussed thus far, the REM system has no particular provisions for security mechanisms over and above those placed on the user-level tools by the underlying operating system. We believe this is the generally correct model. However, we clearly must require access control for external notifications. A natural approach for doing so would be to layer such notifications on top of SSL connections in order to leverage SSL's authentication capabilities. We could potentially augment this with an authorization capability allowing a researcher to define which external monitors can communicate with *remd* and what sort of messages they can send.

3.3   Details

The high-level architecture sketched above is realized through a system whereby each experiment keeps a variable list that can be arbitrarily populated with state information by the experiment configuration and the measurement tools as they are executed. For instance, a measurement tool's argument list can be populated by the configuration setting variables for the tool. Wrapper scripts consult the variable list and add to it information about the outcome of a particular measurement. Once a measurement is finished and the updated list returned, *remd* executes the transitions, which are specified using arbitrary Python code that runs in the context of a given variable list. Using this scheme the *remd* is only required to manage the overall measurement *process* and not have any understanding of the measurements themselves. Thus, *remd* is charged with tasks such as moving variable lists around, executing transition code from the configuration, managing processes, and stopping processes that take too long.

3.4   Interfacing to External Resources

A REM system such as described above would provide a solid foundation for conducting fundamentally new and different measurement studies. However, the system can be more useful still if it were to contain the ability to interact with different types of external resources. Below we sketch two possibilities.

Measurement Infrastructures. Often we can derive more information about a network anomaly by probing the network from multiple vantage points. For instance, a DNS failure may be a local problem to a given network or a more general global problem with one of the root DNS servers. If we perform DNS lookups at only one point in the network (i.e., where *remd* is running), we can fail to observe the full scope of the problem. However, by running the same DNS query from a number of distributed points in the Internet, a more complete story about the failure might emerge. Thus, we should aim to interface the REM system with distributed measurement systems such as *scriptroute* [19] or DipZoom [17]. Such interfaces provide the ability to run reactive measurements at many points in the network simultaneously to gather as much information as possible about network anomalies. Note that by using wrapper scripts, the REM system can accommodate such interfaces without any particular extensions to the general framework: we simply write wrapper scripts invoked by *remd* that, for example, execute *scriptroute* tools to run measurements on alternate hosts and gather the results.

Measurement Repositories. While reactive measurement offers a great deal of power, one deficiency is that sometimes the overt trigger for a failure or anomaly comes *late*: that is, by the time we observe the problem, we may have missed valuable precursors that shed light on the problem's onset. We envision a partial counter to this problem in the form of interfacing to measurement repositories. For instance, wrapper scripts could interface with the bulk packet recorder outlined in [7] in an attempt to try to build understanding about the precursors to some observed phenomena. Another obvious source of information could be the RouteViews repository [11] of advertised routing tables.

4   Summary

Our two major—if preliminary—contributions are (i) developing the general notion of *reactive measurement* as a paradigm that focuses on a measurement *process* as the key to better understanding observed behaviors, and (ii) the design and prototyping of a reactive measurement system to aid researchers in using the technique in their own work. We believe that if the community absorbs and leverages this concept in their experimental designs, it can lead to significant advances in better understanding network behavior. We hope by exposing our initial design to the community we will get feedback on important aspects to include in future versions of our framework.

Acknowledgments

The ideas in this paper have benefited from discussions with a number of people including Fred Baker, Ben Chodroff, Scott Shenker and Randall Stewart. This work was funded in part by Cisco Systems and NSF grants ITR/ANI-0205519 and NSF-0722035. Our thanks to all.

Any opinions, findings, and conclusions or recommendations expressed in this material are those of the authors or originators and do not necessarily reflect the views of the National Science Foundation.

References

1. Allman, M.: Measuring End-to-End Bulk Transfer Capacity. In: ACM SIGCOMM Internet Measurement Workshop (November 2001)
2. Anderson, D.S., Fleizach, C., Savage, S., Voelker, G.M.: Spamscatter: Characterizing Internet Scam Hosting Infrastructure. In: Proceedings of the USENIX Security Symposium (August 2007)
3. Chang, D.-F., Govindan, R., Heidemann, J.: Exploring The Ability of Locating BGP Missing Routes From Multiple Looking Glasses. In: ACM SIGCOMM Network Troubleshooting Workshop (September 2004)
4. Clark, D., Partridge, C., Ramming, J.C., Wroclawksi, J.: A Knowledge Plane for the Internet. In: ACM SIGCOMM Workshop on Future Directions in Network Architecture (August 2003)
5. Huang, H., Jennings III, R., Ruan, Y., Sahoo, R., Sahu, S., Shaikh, A.: PDA: A Tool for Automated Problem Determination. In: Proceedings of USENIX Large Installation System Administration Conference (LISA) (November 2007)
6. Jung, J., Sit, E., Balakrishnan, H., Morris, R.: DNS Performance and the Effectiveness of Caching. In: ACM SIGCOMM Internet Measurement Workshop (November 2001)
7. Kornexl, S., Paxson, V., Dreger, H., Feldmann, A., Sommer, R.: Building a Time Machine for Efficient Recording and Retrieval of High-Volume Network Traffic. In: ACM Internet Measurement Conference (2005)
8. Krishnamurthy, B., Madhyastha, H.V., Spatscheck, O.: ATMEN: A Triggered Network Measurement Infrastructure. In: Proceedings of WWW (May 2005)
9. Mao, Z.M., Rexford, J., Wang, J., Katz, R.: Towards an Accurate AS-Level Traceroute Tool. ACM SIGCOMM (2003)
10. Mathis, M.: Diagnosing Internet Congestion with a Transport Layer Performance Tool. In: Proceedings of INET 1996 (June 1996)
11. University of Oregon RouteViews Project, http://www.routeviews.org
12. Paxson, V.: End-to-End Routing Behavior in the Internet. ACM SIGCOMM (August 1996)
13. Paxson, V.: Bro: A System for Detecting Network Intruders in Real-Time. In: Proceedings of the 7th USENIX Security Symposium (January 1998)
14. Paxson, V., Adams, A., Mathis, M.: Experiences with NIMI. In: Proceedings of Passive and Active Measurement (2000)
15. Paxson, V., Mahdavi, J., Adams, A., Mathis, M.: An Architecture for Large-Scale Internet Measurement. IEEE Communications (1998)
16. Peterson, J.L.: Petri Net Theory and the Modeling of Systems. Prentice Hall, Englewood Cliffs (1981)
17. Rabinovich, M., Triukose, S., Wen, Z., Wang, L.: Dipzoom: the Internet measurements marketplace. In: 9th IEEE Global Internet Symp. (2006)
18. Roesch, M.: Snort: Lightweight intrusion detection for networks. In: Proceedings of USENIX LISA (1999)
19. Spring, N., Wetherall, D., Anderson, T.: Scriptroute: A Public Internet Measurement Facility. In: USENIX Symposium on Internet Technologies and Systems (USITS) (2003)

Towards a High Quality Path-Oriented Network Measurement and Storage System

David Johnson, Daniel Gebhardt, and Jay Lepreau

University of Utah, School of Computing

Abstract. Researchers need current and historical measurements of Internet paths. We built and deployed a complete system designed to fill these needs: a safe, shareable, multi-user active network measurement system probes network paths and reliably records measurements in a storage facility with multiple levels of caching, providing users with fast, flexible querying. Our system, deployed on PlanetLab for over 20 months, has accumulated 940 million measurements and made them publicly available in a separate, federated data repository. Our experience shows that building and running such a valuable research tool poses significant engineering and practical challenges.

1 Introduction

For a multitude of reasons, researchers need current and historical measurements of Internet paths. These reasons include creating or validating network models, using those models to perform experiments under Internet conditions in network testbeds, studying trends and stationarity in network conditions, and selecting Internet paths that tend to exhibit certain properties. We explored one design point on the spectrum of path-oriented network measurement and storage systems, motivated by 1) the needs of a network emulation environment and 2) the need to provide permanent, public repositories of historical measurement data. The result is Flexmon: a shareable, multi-user active measurement system that collects pairwise path data between sites in a network at tunable frequencies, yet protects the network from excess traffic. Its architecture allows multiple clients to schedule their own probes on subsets of nodes in the network, while sharing probe results among clients, amortizing the costs. Furthermore, Flexmon provides a reliable storage path for probe data and stores them permanently in a separate federated data repository.

Flexmon's design provides a measurement infrastructure that is *shared, reliable, safe, adaptive, controllable,* and *accommodates high performance data retrieval.* Each feature is not novel, but the design, the lessons learned, and its eventual more mature implementation should provide an important community resource—and indeed, our accumulated measurements are already publicly available. Flexmon has some features in common with other measurement systems such as S^3 [16] and Scriptroute [13], but is designed to support different goals including shared control over measurements and easy public data availability.

M. Claypool and S. Uhlig (Eds.): PAM 2008, LNCS 4979, pp. 102–111, 2008.
© Springer-Verlag Berlin Heidelberg 2008

Flexmon measurement is controllable and dynamic in that an authenticated user may adjust the measurement frequency of any particular path, while Flexmon itself caps and adjusts the rates based on overall network resources consumed. These features allows higher frequency measurement than what would be possible globally, and thus a more accurate path emulation, while still preserving its safety to large networks like PlanetLab.

Our experience shows that building and running such a system poses major engineering and practical challenges, some of which Flexmon does not yet meet. We have been running Flexmon on the PlanetLab network testbed for over twenty months, accumulating 940 million publicly available measurements between PlanetLab sites. In this paper, we describe the Flexmon architecture and implementation, discuss its reliability, outline our dataset, and draw lessons for developing high quality network measurement and storage systems.

2 Design

In this section, we present Flexmon's architecture and discuss our design choices. We originally built Flexmon to serve as the measurement infrastructure for Flexlab [10]. Flexlab allows researchers to experiment in the controllable, repeatable environment provided by the Emulab [15] testbed, but using link characteristics that are dynamically updated using traffic models derived from a real network, specifically the PlanetLab [8] testbed embedded in the Internet. Consequently, Flexmon inherits several design goals from Flexlab.

2.1 Design Choices

Flexmon's design was driven by four primary requirements: the need to obtain and provide "raw measurements" to researchers, the need to operate within unreliable networks, the desire to obtain data that are useful for many researchers, and the need to protect the network from excess traffic.

First, Flexmon is designed to measure end-to-end network path properties, using (typically standard) external programs wrapped by a script that canonicalizes the parameters. Our initial experiments with packet-pair and packet-train tools, including pathload [3] and pathchirp [9], produced poor results on PlanetLab due to its extremely overloaded hosts. Consequently, we currently use two simple, controllable tools: a modified version of iperf measures bandwidth, and fping measures connectivity and latency. However, this experience reveals the importance of allowing multiple measurement tools. Flexmon's design supports the addition of new measurement tools, as they prove desirable in the future.

The design also supports two methods for running these tools, either *one-shot* or *continuous*. In one-shot mode, the tool is spawned at regular intervals to produce a single measurement that is captured by the system. This mode makes it easy to integrate existing probe tools such as ping and iperf that can produce a "summary" result before exiting. However, this mode can cause high overhead when the probe frequency is high. In continuous mode, the tool is spawned once, for a longer duration, and its periodic results are collected by the path prober. Many probes can benefit from collecting state over an extended period, while still reporting periodic results.

Our system must function over unreliable networks. PlanetLab is a heavily utilized, multi-user overlay testbed with hundreds of nodes distributed around the world. Thus, like the Internet it is a part of, PlanetLab nodes are often unresponsive. To be useful such an environment, Flexmon must be reliable in a number of key ways. When nodes return from an unresponsive state, they must quickly rejoin the measurement network and continue probing. During a network outage, Flexmon nodes must continue probing and reliably store results to persistent storage for delivery when connectivity returns.

Potential users of Flexmon will likely often require the same types of measurements. However, frequent or simultaneous probing with tools such as `iperf` can be both costly and cause self-interference, affecting accuracy. Thus, it is important for Flexmon to allow users to share probe results. In Flexmon, probing costs can be amortized across several users, which can alleviate the larger bandwidth costs incurred by `iperf` and similar network measurement tools.

Finally, both the underlying network and the measurement system itself must be protected from excess measurement traffic. Given our trust model, such traffic is typically caused accidentally, but our experience shows that it is a real problem. Our design can limit global and per-user bandwidth usage, providing both defense in depth and reflecting the reality of traffic limiting on PlanetLab.

2.2 Software Architecture

Flexmon consists of six types of components: *path probers*, a *manager*, *manager clients*, the *auto-manager client*, a *data collection* subsystem, and a federated, permanent *data repository*. Two additional components from the base Emulab system are essential: a reliable but lightweight control system, and a database that includes static and dynamic state of the measurement nodes. Figure 1 shows an overview of the communication between these components.

Flexmon users request that measurement probes be performed among sets of nodes using manager clients. They indicate the nodes to be probed, the type, frequency, and duration of measurements, and any other parameters required

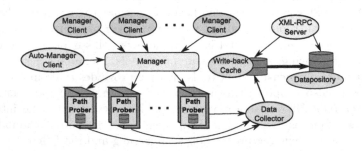

Fig. 1. Flexmon architecture

by the tool. A special type of manager client, the auto-manager client, attempts to maintain a fully connected measurement set for the entire network. The centralized manager receives client probe requests, performs safety checks on them, and forwards the requests to path probers.

Path probers run on each network node and receive control commands from the manager. Commands mirror the client requests, allowing changing of the measurement destination nodes, the type of measurement (e.g., latency), and the measurement frequency and duration.

Path probers send results to a data collection service that caches and batches before sending them on to the permanent federated repository, the Datapository [1]. To speed up both queries and updates, the data collector places the results in a smaller database, which functions as a write-back cache of the measurements taken in the last 24 hours. New data in the cache is flushed hourly to the Datapository database. Finally, an XML-RPC server provides publicly available query functionality to users, allowing queries to its pairwise internal cache, the 24 hour cache, and the Datapository database.

We anticipate that researchers will typically use manager clients to schedule probes at higher frequencies between nodes of specific interest, while the auto-manager client maintains low-frequency, low-cost, network-wide measurements. As an example, the Flexlab [10] testbed uses a manager client to run a high-fidelity path measurement between PlanetLab nodes to determine the initial conditions for link shaping within an emulated network.

Multiple manager clients may send requests specifying different measurement frequencies for the same path. The path prober maintains a queue of probe requests ordered by frequency, and serially executes the highest frequency probe request per type (i.e., available bandwidth estimation, latency, connectivity). Once the duration of that request expires, it is removed from the queue, and the prober executes the next probe. Since the prober runs only the highest frequency probe request per type, all users of the system will see at least the rate of measurements that they requested.

Flexmon can reject probe requests based on resource constraints in the network. For instance, network administrators for probed nodes may request that Flexmon limit its bandwidth consumption to a specific rate. Flexmon maintains a per-path available bandwidth average based on prior available bandwidth probes, and performs admission control on probe requests by computing the expected bandwidth utilization of the probe. If the expectation is that the request will exceed a global limit, or a per-user limit, the probe request would be rejected.

3   Implementation

Flexmon currently serves as the measurement infrastructure for Flexlab, a service running on Emulab. Many of its central components run on Emulab servers, and it measures a subset of the PlanetLab network. In this section, we provide background on how Flexmon is deployed across Emulab and PlanetLab, and discuss implementation details for key system components.

Deployment. Emulab provides a portal [14] to PlanetLab, through which PlanetLab resources can be used in much the same way as other Emulab resources. To deploy Flexmon's path probers on PlanetLab nodes, we create an Emulab experiment containing all PlanetLab nodes considered "live" by Emulab. When this experiment swaps in, Emulab automatically deploys the measurement scripts and tools and starts the node's path prober daemon. Although the number varies over time, Flexmon typically monitors 275–350 PlanetLab nodes .[1]

Background Measurements. From the set of PlanetLab nodes in the Flexmon experiment, the auto-manager client chooses a subset so that each node represents a unique PlanetLab site. The auto-manager client requests all site-pairs latency, connectivity, and bandwidth probes to run with infinite duration between all nodes in this set. The auto-manager client chooses a single node per site because all nodes at one site should exhibit similar path characteristics to nodes at another site. The auto-manager client prioritizes site nodes based on least CPU utilization to minimize the effect of observed latencies in process scheduling on PlanetLab nodes [10, 12], and updates priorities as loads change.

The auto-manager client parameterizes its latency probing by period (each path is measured after an interval), and bandwidth probing by duty cycle (the fraction of time a path prober is measuring bandwidth to any destination). All-sites latency probing is inexpensive when compared to all-sites bandwidth probing, even in a large-scale network such as PlanetLab. However, all-sites bandwidth probing can become extremely costly and could cause Flexmon to run afoul of PlanetLab bandwidth caps. Although Flexmon allows its administrator to set global and per-user caps , it remains important for background probing to leave space under the cap for high frequency, per-user manager client probing. We have found that by setting the latency probing period to 600 seconds, and the bandwidth probing duty cycle to 10%, auto-manager client probes do not exceed PlanetLab caps, and leaves sufficient resources for manager clients.

Probing. A path prober runs on each node in the measurement network. Each path prober receives commands from the manager that control probe execution. Commands result from probe requests to the manager, and set the destination nodes, the probe types to be executed (e.g., latency or bandwidth) between the source and each destination node, the mode in which the probe should run (one-shot or continuous), the period between individual tests, the total duration of probing, and any additional arguments. The path prober spawns wrapper scripts that translate the canonicalized parameters to tool-specific arguments, runs the tool, and converts its results and error conditions into a generic form. For one-shot probes, a result message is sent to the data collector once the probe has finished execution. For continuous probes, the path prober monitors the output of the probe and sends periodic messages to the data collector. Messages contains

---

[1] Regular PlanetLab users will note that this number is lower than the number of nodes typically reported as live by CoMon [6]. Emulab's portal to PlanetLab creates a richer environment on nodes than the default PlanetLab sliver creation method, and thus requires a higher level of node health.

the node pair, probe type, the probe result (including error information), the time at the start of probe execution, and a magic protocol ID.

When deciding how to estimate available bandwidth, we first experimented with several packet-pair and packet-train tools, including `pathload` [3] and `pathchirp` [9]. Others report those two programs to work acceptably well on PlanetLab [5], but in our experience they often returned extremely unreliable results, or none. Therefore, we estimate available bandwidth with `iperf`, a so-called Bulk Transfer Capacity method. `iperf` consumes much bandwidth during tests, but it has the advantage that, by using a real TCP flow, it obtains a highly accurate measurement of the bandwidth seen by a TCP flow, including taking into account the reactivity of other flows. We extended `iperf` with our own iperf daemon, since iperf produced memory leaks during long runtimes. Each path prober runs an iperf daemon to handle incoming probes from other nodes.

Flexmon uses the `fping` utility to measure latency and detect path outages. When a loss occurs, a state machine drives a frequency-adaptive probing process to distinguish packet loss from true connectivity failure (in four seconds), and to subsequently detect connectivity restoration (within ten seconds).

Probing tools may experience errors due to underlying network behavior. Path probers capture certain errors and report them as anomalous measurements since errors provide users with useful information about the state of the network. We currently capture timeout, unresolvable hostname, and host unreachable errors. However, our system can flexibly record arbitrary error conditions.

Flexmon is designed to run safely on unreliable networks. Since PlanetLab is a large, heavily-utilized network, it is inevitable that nodes will periodically reboot or become unresponsive for extended periods of time. If a PlanetLab node is rebooted, or the sliver is reset, the path prober restarts when the sliver does. However, we chose not to have the path prober checkpoint the current running state and resume from it during failure recovery. Each path prober has no knowledge of the overall system goals, and the manager or auto-manager client may have already adapted to deal with the loss of particular nodes.

Measurement Transfer and Storage. Flexmon's reliability and availability are greatly improved by strategic buffering, reliable probe result transfer, and caching. Each path prober sends probe result to the data collector over UDP. Before a prober sends a result, it first inserts it into a Berkeley database, which acts as a stable storage buffer. The prober then sends the message to the data collector and retransmits after five seconds if the result message is not acknowledged by the data collector. The data collector does not acknowledge the result message until it has been successfully inserted into a SQL database; therefore, it must be aware of duplicate measurements, and will drop them when they are detected. Through this mechanism, Flexmon largely ensures application reliability; however, measurements may be lost if a path prober node's disk fails while the data collector is not running.

The data collector maintains the caching database containing measurements taken within the last 24 hours, making result polling and queries on recent data much faster than if all results were inserted directly into the `Datapository`. A script reliably "flushes" measurements in this cache back to the Datapository each hour, ensuring that measurements older than 24 hours are not aged out of the cache until they are successfully entered into the Datapository.

Query Interfaces. To facilitate efficient and easy data access, while minimizing security concerns and increasing functionality, Flexmon provides several interfaces to query measurement result data. First, as mentioned in the previous section, Flexmon maintains the results from the previous 24 hours in a database that acts as a write-back cache. This database, into which the data collector inserts new measurements, provides a reasonably efficient query mechanism for services that require access only to recent data. At this time, the users database is accessible by any researcher with an Emulab account. However, due to security and performance risks, we only provide raw SQL access to the Datapository database to those "power users" who must be able to compose their own queries.

Instead of providing raw SQL access to the Flexmon databases to all potential users, we provide a safer, controlled means of accessing measurement data. Flexmon runs a simple XML-RPC server that periodically polls the users database and keeps an in-memory cache of the single most recent latency and bandwidth measurements for each known path. The XML-RPC server ages measurements out of its internal cache once they reach an age of 24 hours. It also provides a simple API that can be used to query either the users database or the Datapository database, depending on the specified time interval . The getMeasurements method requires an interval to restrict query responses to a manageable size, and can filter results based on source and destination node and site, as well as on basic measurement value constraints. Another method, getFullyConnectedSet, finds a max-clique from the data in the in-memory cache, and can restrict its search to a specific set of nodes.

4   System Status

Flexmon has been monitoring sets of PlanetLab nodes since February 2006, and has placed approximately 940 million measurements of pairwise latency and available bandwidth in the Datapository, accounting for approximately 89% and 11% of total measurements respectively. Until December 2006, Flexmon ran in "beta" mode, and underwent several architectural changes. After this time, although several bugs were fixed, the system remained largely unchanged, aside from occasional changes in the set of PlanetLab nodes monitored.

The number of nodes in the experiment can change over time. When we change the experiment configuration, we normally restart key Flexmon daemons, such as the auto-manager client and the manager. Depending on when restarts occur, there may be slight anomalies for a brief window of time in the measurement archive. Prior to fixing a bug, many of our PlanetLab nodes were given bandwidth caps by PlanetLab for a short time due to detected overuse.

Since Flexmon path probers may not always be able to conduct a pairwise site measurement successfully (i.e., due to unexpected node unavailability, path outages, packet filtering at PlanetLab sites), measurement results stored in the

`Datapository` also include pairwise errors. For instance, over the measurement history, approximately 17% of latency measurements and 11% of bandwidth measurements represent errors. Of known latency errors, approximately 74% are timeouts; 18% are DNS lookup failures; and 6% are ICMP unreachable errors.

5 Metrics

We analyzed a snapshot of logfiles produced by Flexmon during a 100-day period to evaluate key properties of our system. There are times when the number of available sites is much lower due to the inherent unreliability of distributed systems and PlanetLab. The number of nodes in the experiment changes over time. During the period analyzed, we monitored a median of 151 sites.

We first evaluate the number of available sites in our system during the period. We extract the number of available sites from periodic liveness checks performed by the auto-manager client. Figure 2(a) shows the available *sites* over the period. Near Day 23, there is a sharp drop in the available sites, caused by an experiment restart. Overall, this graph demonstrates a relatively stable number of available sites.

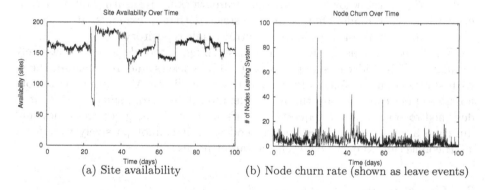

(a) Site availability   (b) Node churn rate (shown as leave events)

Fig. 2. Measurement node availability and churn rate

We also analyzed the churn rate of nodes in our system. We compute the churn rate as the number of nodes that "left" (were newly unresponsive) in a single period of liveness checks. Figure 2(b) shows the churn rate over the period. First, we see large spikes near Day 23 that are caused by the experiment restart, where a large number of nodes suddenly left the system. These correlate well with the decreased site availability shown in Figure 2(a), as do other spikes in this graph, which may represent transient failures.

6 Related Work

Researchers have built a number of network measurement infrastructures, each with its unique spin on the measurement task.

Scriptroute [13] provides a safe, publicly-available network probe execution environment. Users submit probe scripts written in an interpreted language to Scriptroute servers, which execute the scripts while ensuring that scripts do not exceed resource limits and do not send malformed packets. Flexmon also constrains probes according to resource limits. However, since Flexmon prevents users from running arbitrary tools or algorithms, and only allows probes between nodes within the monitored network, a malformed packet filtering mechanism such as that provided by Scriptroute is unnecessary. Due to its more permissive trust model, Scriptroute does not allow node-local data storage, which compromises reliability. It is also not linked to a public data repository.

ANEMOS [2] is an extensible measurement system which stores results in a central database. However, task scheduling is done on the centralized *Coordinator* rather than in a distributed fashion, and results are not buffered at the *Workers*. These two traits limit its usefulness on an unreliable network. While Flexmon demonstrates scalability at least to hundreds of nodes, ANEMOS has only been tested to tens. The work in ATMEN [4] describes an open framework for providing network measurement services and querying the results. Its distributed architecture provides scalability for measurement, but does not provide functionality for a general user or application to access the entire history of all collected data.

NIMI [7] is a well-known measurement framework for measuring Internet traffic, with a trust model that is more restricted than Flexmon's. It is secure, scalable, and extensible, but lacks a central result repository.

TCP Sidecar [11] is the foundation of sideping, a tool designed to passively estimate RTTs. Sidecar snoops on sent TCP packets and retransmits them, with subtle changes. When Sidecar receives a duplicate ACK from the remote host, sideping can estimate the RTT. Flexmon differs from sideping since it is designed to service probe requests from its users with the given frequency and duration arguments, between specific nodes, so it cannot passively wait for a TCP connection to occur between the target nodes.

7   Conclusion

We have described Flexmon, a shareable, path-oriented, active network measurement system with reliable measurement storage. Our experience shows that building such a system poses significant engineering and practical challenges. It should ensure reliable measurement storage, avoid overloading monitored nodes and networks, and function in unreliable networks with heavily loaded nodes.

It will require a major effort to provide such a system that is truly reliable, safe, and efficient, with high availability and high performance. The level of effort is probably similar to the effort we ourselves expended in building the initial version of the Emulab network emulation testbed. However, Flexmon is a real, working system that has collected 940 million measurements, it has been successfully used by the Flexlab network testbed, and we plan to evolve it to a permanent and production-quality measurement and storage system.

Acknowledgments

We thank many of our colleagues in the Flux Research Group for their significant contributions. Robert Ricci, Mike Hibler, Leigh Stoller, and Sachin Goyal added functionality to Flexmon. Robert, Sachin, and Kirk Webb also helped operate the deployed system. Pramod Sanaga and Kevin Atkinson assisted in gathering performance numbers. Eric Eide provided feedback and editing support. We thank David Andersen for his work on the Datapository.

References

[1] Andersen, D.G., Feamster, N.: Challenges and Opportunities in Internet Data Mining. Technical Report CMU–PDL–06–102, CMU Parallel Data Laboratory (January 2006), http://www.datapository.net/

[2] Danalis, A., Dovrolis, C.: ANEMOS: An Autonomous Network Monitoring System. In: Proc. PAM, San Diego, CA (April 2003)

[3] Jain, M., Dovrolis, C.: End-to-End Available Bandwidth: Measurement Methodology, Dynamics, and Relation with TCP Throughput. IEEE/ACM Trans. Networking 11(4), 537–549 (2003)

[4] Krishnamurthy, B., Madhyastha, H.V., Spatscheck, O.: ATMEN: A Triggered Network Measurement Infrastructure. In: Proc. WWW (May 2005)

[5] Lee, S.-J., et al.: Measuring Bandwidth Between PlanetLab Nodes. In: Proc. PAM (March–April 2005)

[6] Park, K., Pai, V.: CoMon: A Mostly-Scalable Monitoring System for PlanetLab. OSR 40(1), 65–74 (2006)

[7] Paxson, V., Mahdavi, J., Adams, A., Mathis, M.: An Architecture for Large-Scale Internet Measurement. IEEE Comm. 36(8), 48–54 (1998)

[8] Peterson, L., Anderson, T., Culler, D., Roscoe, T.: A Blueprint for Introducing Disruptive Technology into the Internet. In: Proc. HotNets-I (October 2002)

[9] Ribeiro, V., Riedi, R., Baraniuk, R., Navratil, J., Cottrell, L.: pathChirp: Efficient Available Bandwidth Estimation for Network Paths. In: Proc. PAM (April 2003)

[10] Ricci, R., et al.: The Flexlab Approach to Realistic Evaluation of Networked Systems. In: Proc. NSDI (April 2007)

[11] Sherwood, R., Spring, N.: A Platform for Unobtrusive Measurements on Planet-Lab. In: Proc. of WORLDS 2006, Seattle, WA (November 2006)

[12] Sommers, J., Barford, P.: An Active Measurement System for Shared Environments. In: Proc. IMC (October 2007)

[13] Spring, N., Wetherall, D., Anderson, T.: Scriptroute: A Public Internet Measurement Facility. In: Proc. USITS (March 2003)

[14] Webb, K., et al.: Implementing the Emulab-PlanetLab Portal: Experience and Lessons Learned. In: Proc. WORLDS (December 2004)

[15] White, B., et al.: An Integrated Experimental Environment for Distributed Systems and Networks. In: Proc. OSDI (December 2002)

[16] Yalagandula, P., et al.: S3: A Scalable Sensing Service for Monitoring Large Networked Systems. In: Proc. Workshop on Internet Network Mgmt (September 2006)

On Community-Oriented Internet Measurement

Mark Allman[1], Lann Martin[2], Michael Rabinovich[2], and Kenneth Atchinson[3]

[1] International Computer Science Institute
[2] Case Western Reserve University
[3] Baldwin-Wallace College

Abstract. In this paper we describe a new measurement framework that researchers can use to abstract away some of the mundane logistic details that tend to dog every measurement project. The measurement community has outlined the need for better ways to gather assessments from a multitude of vantage points and our system is designed to be an *open community-oriented* response to this desire. While many previous efforts have approached this problem with heavyweight systems that ultimately fizzle due to logistical issues (*e.g.*, hosts breaking and no money to replace them) we take the opposite approach and attempt to use the lightest possible weight framework that allows researchers to get their work done. In particular, we take the approach of designing a system without any sort of central "core" component and therefore the system has no single point of failure. In addition, our proposed system is *community-oriented* in that there is no central control and we build just enough mechanism for the community to get their work done and police the infrastructure. In addition, our proposed system works in an open fashion such that results from the community's infrastructure are immediately provided to the community through publicly available "live feeds".

1 Introduction

The Internet has become a vastly complex and heterogeneous system that defies simple characterization or measurement. Researchers gain fundamental understanding from detailed measurements spanning a wide variety of vantage points around the network. A thriving sub-community of networking researchers has emerged that focuses on Internet measurement and analysis. Arguably, this sub-community has greatly enhanced global understanding of a wide variety of aspects of how networks work "in the wild" (*e.g.*, operations of the routing system, better understanding of peer-to-peer transfers, how various attacks operate, etc.). With this understanding come new and better techniques for designing and deploying Internet technologies. Our goal is to both enhance this sub-community's ability to provide further understanding, as well as enhance the entire community's ability to assess the efficacy of new ideas through live measurements.

The research community has clearly stated its need for more and better measurement data. An NSF-sponsored workshop on "Community-Oriented Network Measurement Infrastructure" brought together a set of measurement experts who noted a variety of community needs [5]. Among the needs articulated were both the need to more easily run large-scale Internet measurements and the need for datasets from a broad range of networks. In this paper we provide an initial sketch of a system that addresses both of these desires.

M. Claypool and S. Uhlig (Eds.): PAM 2008, LNCS 4979, pp. 112–121, 2008.
© Springer-Verlag Berlin Heidelberg 2008

Internet measurement studies often fall into one of two camps: (i) those that require researchers to expend large amounts of time on a formidable amount of mundane logistical details in order to run their measurement tools and collect data from a variety of locations and (ii) small-scale studies that only consider small pockets of the network and therefore may not be indicative of large-scale behavior. Our goal is to provide a *community-based* measurement framework to address some of the problems associated with large-scale measurement. We intend to form a *lightweight* measurement platform that maintains *no* dedicated infrastructure. Instead, it relies on using a distributed hash table (DHT) (*e.g.*, OpenDHT [9], an overlay substrate used by a variety of other applications) to handle all communication needs for a mesh of measurement hosts. We will provide tools and libraries to aid in the communication tasks specifically required to undertake Internet measurements (*e.g.*, find measurement points, form measurement requests and collect results). Our over-arching goals are to ease the pain involved with conducting large-scale measurement studies such that researchers can both (a) spend more time focused on gaining insight about the network and how their new technologies work and less time on logistics and (b) have better access to large-scale infrastructure and data such that researcher will be incentivized to move away from small-scale studies. The Internet has benefited from community effort for a number of innovations. Community members proved willing to contribute resources to projects of individual research groups, such as seti@home, traceroute@home and DIMES. We expect that they would be even more willing to take part in an effort that benefits the entire community.

Abstracting the mundane details of measurement away from researchers should not be taken as a small contribution. In fact, our experience is that much of the effort associated with large measurement studies is spent getting the mundane logistics right. While we do not provide a framework to rid researchers of all the logistical headaches of network measurement we provide a framework that takes care of a number of the issues. With this in mind we sketch several aspects of our framework:

- The infrastructure can support a fluid set of measurement points that are provided and administered by the community. Unlike other efforts there is no central management required, making this system a truly community-oriented effort that is *of* and *for* the network research community.
- Our envisioned system uses a general-purpose DHT for the "glue" that (loosely) connects the system components. The lack of a central core is a feature in that no central maintenance is required and no single point of failure exists.
- Researchers will be freed from many, but not all, of the logistic details of recruiting measurement points to focus their attention on the important details of the measurements themselves (techniques, data analysis, etc.).
- The system requires no centralized maintenance beyond keeping the DHT running. We envision that the research community will keep a DHT such as OpenDHT running for a variety of purposes anyway and so using the DHT to coordinate measurements is not an extra burden. If this is not the case and yet the community still desired such a system for measurement purposes a DHT can be readily built from existing and available DHT software.
- Long-running measurements that benefit the entire community can be run with community-wide resources. For instance, a setup similar to CAIDA's *skitter*

system [4] could be built and be supported by a distributed set of organizations—none of which control, or can hinder (*e.g.*, due to funding or manpower issues), the overall data collection operation.

- Small and focused sets of measurements can be *easily* taken between a consenting group of measurement points. That is, the group of measurement points used in a particular experiment may be organized specifically for that experiment and not assembled from generic measurement hosts donated to the community. This can help with measurements that are too unknown or specialized to run on shared measurement points (*e.g.*, due to security concerns or because an experiment requires a specialized kernel).

- Since the measurement results are reported through the DHT, anyone can pull down the results of measurements as they are completed. This allows the entire community to benefit from measurements involving the community's shared infrastructure immediately, rather than waiting for the raw data to be posted to some measurement repository and indexed in systems like DatCat [3] or PREDICT [1]. (Note, as discussed in § 3 our system provides immediate but *short-term* storage of results. Therefore, archiving and indexing measurement results in long-term repositories is orthogonal to our framework.)

- The "barrier to entry" for doing large-scale Internet measurement studies is quite high due to the need for a distributed set of measurement points and the time required to coordinate measurements and observations. This shuts many researchers with small-to-modest resources out of the entire area of research (or, relegates them to conducting limited studies, as noted above). Our proposed system will open the field of sound, large-scale Internet measurement to a much broader community of researchers than are currently engaged in this field.

- In addition, having a lightweight measurement infrastructure that can be easily used can encourage researchers who are not engaged in "Internet measurement" per se to both (*i*) test their ideas out on the real network and (*ii*) take broad measurements to solidly ground their work in the actual operation of the network.

- While we are proposing an "open" infrastructure the security implications of nefarious use of the platform must be taken into account. We discuss mitigating such concerns in § 3.2.

2   Related Work

Our work is related to two classes of previous efforts: (*i*) measurement taking infrastructures and (*ii*) data dissemination systems.

A number of measurement taking infrastructures have been developed, each with their own wrinkle (*e.g.*, NIMI [7], Surveyor [6]). Generally these systems have been more heavyweight than the system we propose. These systems have features that we do not include in our design, such as allowing for the updating of tools, coordinating measurements, stronger and more fine-grained notions of access control, etc. Our system is in some sense on the opposite end of the spectrum—making up for a lack of features by making the *key* tasks as easy as possible. In addition, we note that in many cases these heavyweight all-encompassing measurement infrastructures have ultimately required more upkeep than their designers and operators could handle (*e.g.*, due to the cost of replacing worn out portions of the infrastructure) and so have withered.[1] Again, we take the opposite approach and focus on designing a framework that can be used without any sort of central authority and without relying on any particular organization other than the community at-large to maintain infrastructure. Similarly, the current DipZoom project [8] uses a peer-to-peer approach, but aims to leverage Internet users at large as measurement providers and uses a central core that must be maintained. This approach distributes the cost and effort to maintain the entire system.

Taking measurements is only one part of our system. Since we are using an open DHT for all communication, the results of the measurements can be retrieved directly from the DHT by the community at-large.[2] These "live feeds" of data then benefit the entire community. A number of efforts provide access to archived measurement data (*e.g.*, as indexed in CAIDA's Data Catalog [3]). While our system provides direct access to data without such a catalog, the systems are actually orthogonal. We do not envision keeping measurement results in the DHT indefinitely. Rather, we envision that the data will age out on the order of days after it was produced. Therefore, while the community can latch on to live feeds, longer term archival and indexing systems will still be required.

3 System Architecture

As outlined above, the proposed measurement system is centered around an open distributed hash table such as OpenDHT [9]. Our only requirement for the DHT is that it support a *get()/put()* interface. That is, *put (k,value,t)* places *value* into the DHT under hash key k with a time-to-live of t. Note that multiple values can be placed in the DHT for a given key. The DHT is queried using *get (k)* to retrieve all the values stored under the hash key k.

Fundamentally, there are three types of actors and three operations for a measurement system. The actors consist of (*i*) measurement requesters who desire some assessment of the network, (*ii*) measurement points (MPs) that provide certain types of measurements upon request and (*iii*) so-called "watchers" that do not request measurements, but do track the "live feeds" by retrieving measurement results from measurements scheduled by others. The operations that must be supported are: (*a*) identifying a remote measurement point suitable to provide the desired measurement, (*b*) requesting a measurement be conducted by a remote measurement point and (*c*) retrieving measurement results when available. In the following subsections we discuss in detail how the system works and several additional considerations.

3.1 Tables

Since we employ a DHT to loosely couple all the entities in our system, all communications happen through entries in various tables held in the DHT. The various actors in

---

[1] Note that not all infrastructures have met this fate. For instance, the *skitter* infrastructure [4] (and its descendant *archipelago* [2]) has been kept running for close to 10 years (through much hard work).

[2] Clearly, a researcher could encrypt measurement results before placing them in the DHT to prevent community access, but this runs counter to the spirit of the system.

the system are responsible for inserting new table entries, maintaining existing entries and polling the DHT tables periodically to find new entries. Our system does not call for the explicit removal of items from the DHT, but rather assumes they will be aged out (based on the time-to-live described above.[3] We now discuss the three basic operations provided by the platform.

Identifying Measurement Points. The first key task for researchers wishing to make use of our system is finding the names of the tables to deposit measurement requests into and finding the names of the tables that can be monitored for results. Methods that would accomplish this goal depend on the usage scenarios of our proposed framework. If the experimenter is simply using the framework to interact with a set of well-known nodes that have been constructed for a particular study then the task of finding MPs is unnecessary. However, if a researcher wishes to make use of a set of community-provided MPs (*e.g.,* hosts setup to run *wget* on request) then some discovery process needs to be put in place. Our system contains a master table *AllMPs* that includes information about each measurement provider. At a minimum this master list will indicate the measurement type including version (*traceroute*, *wget*, etc.), acceptable arguments, the name of the DHT key monitored for measurement requests and the name of the DHT key under which results will be deposited. In addition, ancillary information may also be given (*e.g.,* tool version, operating system and version, location of the measurement host, etc.). The entries in the master table are populated and maintained by the MPs as they come online. If an MP becomes inactive, its entry in the master table will age out, so MP failures may only cause some number of measurements requests to go unfulfilled—already possibility due to the best-effort nature of our system. Also note if some host provides multiple measurement types (*e.g.,* ping and *pathload*) then it will have multiple entries in the master list. That is, each entry in the list is scoped to one measurement type.

Requesting Measurements. Requesting a measurement involves simply inserting an appropriate record into a table that a given measurement point regularly consults—as determined, for instance, by consulting the master list of measurement providers discussed above. The time-to-live of measurement requests should be fairly short (minutes) since MPs are assumed to be polling for new requests regularly. Each request will give the time the measurement point should run the measurement[4], the arguments to run the tool with and the name of a DHT table to place the results into (in addition to the table where all a given MP's results are deposited).

Reporting Measurement Results. Similar to issuing a measurement request, reporting measurement results involves putting the data into the DHT and then placing at least one pointer to that result into appropriate tables. First, each measurement result is put into the DHT under a unique hash key, U. For instance, a Universal Unique Identifier (UUID) could be used for U (as returned by *uuidgen* or similar). Using a unique

---

[3] Note that OpenDHT has a built-in TTL limit to deal with overly long TTL requests. A TTL limited coupled with a web-of-trust (see § 3.2), mitigates the potential clogging problems that would result from overly long TTL values.

[4] MPs will be expected to be roughly time synchronized (*e.g.,* to within seconds, not minutes). This could be tracked with a heartbeat measurement built on top of the generic platform.

identifier for each measurement both avoids name clashes between MPs and allows the measurement results to reside in the DHT only once, but be indexed in a variety of ways. The key U will be placed into both the results table given in the measurement request and the results table advertised by the measurement point as the depository for all its results.[5] Additional pointers could be placed in other tables as the measurement point deems appropriate (*e.g.*, a table for all *ping* measurements taken in Europe).

We note that DHTs often have a limitation on the size of each entry. For instance, OpenDHT has a 1024 byte limit on the size of the records that can be placed into the system. Obviously, this may be inadequate for many measurement results and therefore the MPs will have to split the results across a number of entries with the consumers of those results being required to reassemble the pieces. As discussed in § 3.3 we intend to make this process seamless for MPs and measurement consumers by providing fragmentation and reassembly primitives. Therefore, instead of using the DHT's standard *put()* and *get()* functions, alternate forms will be available that abstract away any required fragmentation and reassembly.

3.1.1 Example. We now step through an example usage of our framework. This example is meant to be illustrative and help the reader gain intuition in the system, rather than exhaustively showing all possible behavior and capabilities.

MP Registration. When a measurement point for a particular measurement comes on line it registers four pieces of information in the "AllMPs" master table: (i) the type of measurement and version being provided (*e.g., ping-0.45b*), (ii) the name of the request queue the measurement point services (*e.g.,* "reqQ"), (iii) the name of the list that the measurement point adds results to upon completion (*e.g.,* "respQ") and (iv) other ancillary data that may aid researchers (*e.g.,* location of measurement point, operating system version, etc.).[6] Example:

  put ("AllMPs","ping-0.45b reqQ respQ extra_info")

Finding MPs. When a researcher wants to run a particular kind of measurement they can access the "AllMPs" table to obtain a list of the MPs, their capabilities and the tables they use. Example:

  get ("AllMPs")
  $\Rightarrow$
  ping-0.45b reqQ respQ extra

Measurement Request. After a researcher has determined MPs that meet their needs they request a particular MP perform a measurement by adding an entry to the MP's request queue that gives (i) the time the measurement should be undertaken (*e.g.,* 184866301), (ii) the name of a result queue the researcher will monitor for results (*e.g.,* "MyResults") and (iii) arguments to the particular measurement tool (*e.g.,* "-n www.icir.org"). Examples:

---

[5] Note that there will inevitably be additional details included with the results, such as a checksum of the results, time the measurement was taken, etc. These details are omitted here, where we focus on the high-level design.

[6] Note: We have not yet added structure to this information, but such structure (or partial structure) would likely be needed to make this field useful for automated processing.

put ("reqQ","184866301 MyResults -n www.icir.org")
put ("reqQ","184866601 MyResults -n www.icir.org")

Measurement Point Polling. Periodically, the measurement point polls the DHT to retrieve its request queue. If the MP sketched above polled it would find the two measurements inserted into the queue in the last step. Example:

get ("reqQ")
⇒
1184866301 MyResults -n www.icir.org
1184866601 MyResults -n www.icir.org

Running Measurements. Upon receiving requests the MP schedules and executes the requested measurements. Upon completion assume the results will be held in some local variable **R**. The MP will generate a universally unique identifier as the key under which to place the measurement result (*e.g.*, **U**). After having placed the results in the DHT the MP then places pointers to the results in its own result queue and the queue requested in the researcher's measurement request. Example:

put (U,R)
put ("respQ",U)
put ("MyResults",U)

Researcher Retrieving Results. The researcher who requests some measurements simply polls on the result queue provided in their request to retrieve pointers to measurement results. Following these pointers will then yield the results. Example:

get ("MyResults")
⇒
U
get (U)
⇒
measurement results (**R**, in this case)

Watcher Retrieving Results. An uninvolved researcher can simply watch results roll into the DHT based on other's requests. In order to do this the watcher will first have to identify MPs conducting desirable measurements (as shown above in the "Finding MPs" step. From this information the passive observer can then poll on the MP's response queue for pointers to measurement results as shown in the previous example above (but starting with retrieving the "respQ" table instead of "MyResults").

3.2 Security

As sketched above, the system has a number of security vulnerabilities. First, a measurement requester can attempt to increase the load on a measurement point simply by requesting large quantities of measurements. Even more problematic is the distributed nature of the system which could allow a requester to coax many MPs to simultaneously send (potentially large volumes of) traffic towards a particular victim. Finally, an attacker could launder requests through the measurement infrastructure in an attempt to gain a layer of anonymity. We offer several approaches to mitigate such problems.

First, we note that measurement requests are just that: *requests*. We make no assumptions that the MPs *must* satisfy all (or any) requests. The requests will receive "best effort"-like service. That is, a measurement point should do its best to conduct the requested measurement at the requested time, but does not make any guarantees. Given this notion, every measurement point can implement local policy related to its willingness to conduct measurements to mitigate some of the security concerns. For instance, a measurement point can both limit the rate of requests that will be serviced (in the aggregate and from a given requester) and can monitor and limit the host and network resources a particular measurement consumes—terminating the measurement if certain thresholds are eclipsed (a la ScriptRoute [10] and DipZoom [8]).

Protecting against nefarious use of a given measurement point is difficult within our framework because we do not have a central authority through which requests can be vetted. For instance, an attacker could coax a large number of measurement points to engage in a DDoS of some service. Or, an attacker could launder their web connections through such a service to add a layer of anonymity. The MPs themselves each only understand a small part of an attack which could look nothing at all like an attack from their viewpoint. Rather than trying to somehow vet all requests that are inserted into the system in a centralized fashion, we again take a community-based approach to the problem and offer two mitigations.

- MPs could inform each other about the measurements they are conducting. For instance, a measurement point executing a measurement towards some target host *H* could insert that fact into a table in the DHT. Before MPs run measurements they consult this target-based table to assess the load already being placed on the target before deciding whether to execute the given measurement.
- A second mechanism is that we impose the requirement that all entries placed in the DHT be cryptographically signed. We then construct a table in the DHT whereby researchers can recognize each other's cryptographic keys as legitimate (i.e., to build a web-of-trust). MPs can then only act on requests from known well-intentioned researchers. There is a one-time cost in getting on such a list, but the cost is small (getting a small number of colleagues to vouch for you in the system). This web-of-trust provides a reasonable sense of a requester's intention before running a measurement and accountability afterwards.

While the general problem of trusting requesters in an open measurement system is difficult, our intention is to leverage the fact that when scoped to a system by and for researchers the problem becomes tractable to suitably mitigate with simple techniques.

3.3 Primitives

Our system does not attempt to provide a stock measurement system that will satisfy all researchers and all tasks. Rather, researchers will have to integrate new tools and new data collection techniques into the system as they are needed. The following primitives are designed to aid researchers in this integration task by providing high-level abstractions to the low-level details required to interact with the DHT. In addition, we will provide tools for common tasks that use these primitives.

Registration. As noted above, MPs will register and maintain their presence and information about the measurement tools they provide. While the specific contents of the registration will be tool specific, the process will be common across tools.

Removing Duplicates. Since we rely on polling and on entries in the DHT to simply time out there must be a way for actors to discover entries that have already been processed when retrieving a table. For instance, a measurement point would only want to schedule one measurement no matter how many times a given request is retrieved. Also, retrieving a measurement result once is sufficient and just because a result pointer is observed multiple times does not mean the result needs to be fetched multiple times. Our design calls for an abstraction that only exposes previously unseen items.

Assessing Trust. As sketched above, one common task across measurement types will be interacting with a web-of-trust to assess a requester's legitimacy. Primitives to aid in this process will be key to making such a trust model work.

Fragmentation and Reassembly. As discussed above, measurement results may need to be fragmented across a number of DHT entries due to limitations on the size of a single entry (*e.g.,* 1024 bytes in OpenDHT). Therefore, primitives for fragmentation and reassembly will be required such that researchers and developers can be provided with an abstraction that works on entire measurements and are not bothered by the details of how they are placed into the DHT.

Miscellaneous Tasks. There are important common measurement-oriented primitives that will aid researchers in setting up their measurements. For instance, a common task is to derive a measurement schedule, and primitives for this task that will allow for direct use with the overall measurement framework will be crucial. Another example is for a measurement point to implement an event loop that polls the DHT for needed information and executes measurements at the appointed times. Having a primitive for easily constructing such an event loop will inevitably aid those integrating new measurement tools into the system.

The above primitives are designed to work across a variety of measurement applications. We note, however, that the above list is likely incomplete. As we progress beyond our proof-of-concept implementation (see § 4) we will likely find additional primitives that are broadly useful and we will include these in the released toolkit.

3.4 Passive Measurements

We note that our discussion above is in terms of active measurements. However, passive monitors can also be used in our system. We envision these manifesting themselves in two forms: by-request monitoring and continuous monitoring. In the first category a researcher can place a request into an appropriate DHT table to have a measurement point monitor some facet of the network for some prescribed amount of time. For example, an MP can register as being capable of monitoring a local Web site, and a request could be to watch a web log for the next 10 minutes. The latter category allows for passive monitors to simply run continuously and dump their results into the DHT for public consumption via live feeds (*e.g.,* a distributed dark address space monitor that provides a wide view of malicious activity). As with sharing of any passive measurements the provider may wish to apply anonymization and sanitization policies to the data before release. For instance, a passive monitor might provide the length (in bytes or seconds) of each TCP connection without providing the IP addresses (even in anonymized form).

4   Summary

Because of the established difficulties in maintaining coherent measurement infrastructures, we propose to build a measurement platform with *no* dedicated infrastructure at all. Instead, we utilize an *existing* overlay substrate *already maintained* for a variety of other purposes, and we concentrate all functionality that is specific to our platform in the end hosts. We further make end hosts totally autonomous and loosely connected to the platform: they can join and depart at will without any reconfiguration in the rest of the platform. This allows the platform to grow and shrink naturally with the needs of the community and be resilient to failures (either technical or logistical). It is important to note that we do not tackle all the hard problems associated with measurement, but rather provide a reasonable platform as a basis.

We have built a small prototype of our system that includes a generic client and an MP that provides *traceroute* measurements on request. While modest, this small prototype is aiding us as we flesh out the details of the system as we work towards providing a toolkit for the broader community.

Acknowledgments

We thank Ethan Blanton, Josh Blanton and Yaohan Chen for discussions of the system described in this paper. Vern Paxson and the anonymous reviewers provided valuable suggestions on a draft of this paper. This work was sponsored by NSF grants ITR/ANI-0205519, NSF-0722035 and NSF/CNS-0721890 for which we are grateful.

References

1. PREDICT: Protected Repository for the Defense of Infrastructure Against Cyber Threats, http://www.predict.org
2. CAIDA. Archipelago measurement infrastructure, http://www.caida.org/projects/ark/
3. CAIDA. Internet Measurement Data Catalog, http://www.datcat.org
4. CAIDA. Skitter, http://www.caida.org/tools/measurments/skitter/
5. claffy, k., Crovella, M., Friedman, T., Shannon, C., Spring, N.: Community-Oriented Network Measurement Infrastructure (CONMI) Workshop Report. ACM Computer Communication Review 36(2), 41–48 (2006)
6. Kalidindi, S., Zekauskas, M.J.: Surveyor: An infrastructure for internet performance measurements. In: INET 1999 (1999)
7. Paxson, V., Mahdavi, J., Adams, A., Mathis, M.: An architecture for large-scale internet measurements. IEEE Communications 36(8), 48–54 (1998)
8. Rabinovich, M., Triukose, S., Wen, Z., Wang, L.: Dipzoom: the internet measurements marketplace. In: 9th IEEE Global Internet Symp. (2006)
9. Rhea, S., Godfrey, B., Karp, B., Kubiatowicz, J., Ratnasamy, S., Shenker, S., Stoica, I., Yu, H.: OpenDHT: A Public DHT Service and Its Uses. In: SIGCOMM (2005)
10. Spring, N., Wetherall, D., Anderson, T.: Scriptroute: A public internet measurement facility. In: Usenix Symp. on Internet Technologies and Systems (2003)

On the Effectiveness of Switched Beam
Antennas in Indoor Environments

Marc Blanco[2], Ravi Kokku[1], Kishore Ramachandran[3], Sampath Rangarajan[1],
and Karthik Sundaresan[1]

[1] NEC Laboratories America, Princeton, NJ
[2] Rice University, Houston, TX
[3] Rutgers University, New Brunswick, NJ

Abstract. Switched beam antennas are an attractive extension to indoor wireless LANs due to their increased signal gain in a chosen direction; the gain can be exploited for improving wireless link quality, node
localization and increasing spatial reuse. However, indoor environments
are susceptible to multipath reflections that may reduce the degree of
directionality of the antennas. To this end, in this paper, we address
the following questions that have not been explored well in the open
literature: how directional in reality is a beam with a switched beam
antenna in a reflection-rich environment, and what are the implications
of the observed directionality on spatial reuse and node localization?
And how does the directionality get affected with the characteristics of a
beam such as main and side lobe width, and front to side lobe ratio? We
present results of measurements in a real office setting with a switched
beam antenna built out of an 8-element phase array.

1   Introduction

Several research works demonstrate the benefits of directional antennas in wireless networks, especially for better link quality and spatial reuse [7,10,12,13,14,
16,17], localization [11,15] and security [5]. All these works exploit the ability of a
directional antenna to focus the transmission energy in a particular direction and
suppress the energy in unwanted directions; the ability is often loosely termed
as the *directionality* of the antenna. Most of these works assume environments
such as outdoors where the antenna provides close to the desired directionality,
and presented analytical, simulation and a few prototype [10,13,14] studies to
demonstrate the benefits. In the context of indoor environments, conventional
wisdom appears to be that the benefits of directional antennas may not be as
dramatic as outdoors due to multipath reflections.

Two recent trends, however, motivate renewed interest in using directional
antennas in indoor environments. First, with the recent popularity of enterprise
WLANs for diverse applications such as VOIP, mainstream office applications,
video conferencing and streaming, interest is increasing in the industry for tap-
ping the benefits of directional antennas in indoor enterprise WLANs (e.g. See
Ruckus Wireless [3]). Second, the technology for achieving directionality with

M. Claypool and S. Uhlig (Eds.): PAM 2008, LNCS 4979, pp. 122–131, 2008.

antennas is becoming cheaper and readily available to make it attractive for incorporating into WLAN products easily [1, 3]. Depending on the features and flexibility provided, directional antennas can be classified into patch and sectorized antennas, switched beam antennas, and adaptive beamforming antennas. While patch and sectorized antennas are designed to focus the antenna beam pattern in one fixed direction, switched beam antennas provide several fixed beams out of which one is chosen for transmission and reception, and adaptive beam antennas adapt beams dynamically in signal space to minimize interference to as many other nodes in the network as possible.

In this paper, we focus on studying switched beam antennas in indoor environments. They provide a good tradeoff among the available antenna technologies; they are less bulky than a collection of patch antennas for providing the same amount of network coverage, and they are simpler to implement and incorporate than adaptive beamforming antennas that require significant channel feedback from receivers for forming appropriate beams *dynamically* at a transmitter. With respect to switched-beam antennas, this paper answers through measurements in a realistic setting a set of basic and important questions: *How directional in reality is a beam with a switched beam antenna in a reflection-rich environment? How does the directionality get affected with the characteristics of a beam such as main and side lobe width, front to side lobe ratio, and location of clients?* The only research effort that we are familiar of in the indoor context is [6] that employs 10° beams and focuses on improving link quality; it does not address the more generic questions we ask.

The ability of switched-beam antennas to form directional beams that suppress energy in several directions contributes to increased simultaneous transmissions in the network, often referred to as *spatial reuse*. Further, the increased signal strength in the main beam direction towards the clients, helps *localize* the client within the angular width of a beam. We study the degree of directionality offered by switched-beam antennas in the context of spatial reuse and localization. In particular, we make the following key observations. (1) The notion of directionality is different for different applications such as spatial reuse and localization, and hence the traditional approach of using "gain over an omnidirectional antenna" to quantify directionality [14] is not comprehensive. (2) Although reflections in indoor environments increase the interference in more directions than ideal, there can be several locations in an indoor environment where a directional beam indeed suppresses interference, thereby making spatial reuse possible. (3) For localization, while most of the clients get localized correctly, a few clients get wrongly localized mainly because of a *small* difference in RSSI (Received signal strength) between the best beam for a client and the beam in the actual direction of the client., thereby necessitating intelligent beam resolution mechanisms. (4) Finally, while our experiments verify that thinner beamwidths do yield greater directionality, they do not completely eliminate the impact of indoor reflections, thereby reinforcing the importance of the above implications in indoor environments even with thin beams.

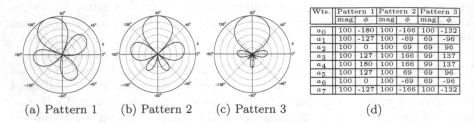

Wts.	Pattern 1		Pattern 2		Pattern 3	
	mag	ϕ	mag	ϕ	mag	ϕ
a_0	100	-180	100	-166	100	-132
a_1	100	-127	100	-69	69	-96
a_2	100	0	100	69	69	96
a_3	100	127	100	166	99	137
a_4	100	180	100	166	99	137
a_5	100	127	100	69	69	96
a_6	100	0	100	-69	69	-96
a_7	100	-127	100	-166	100	-132

(a) Pattern 1   (b) Pattern 2   (c) Pattern 3   (d)

Fig. 1. Antenna beam patterns (a, b, c) and the corresponding element weights (d)

The rest of the paper is organized as follows. Section 2 describes the basics of switched beam antennas. Section 3 presents the experimental methodology that we use to study the directionality of a switched beam antenna in a typical indoor environment. Section 4 presents results and our interpretations from our experiments. Section 5 summarizes the paper.

2   Background

A common way of realizing switched beam or adaptive beamforming antennas is by using phase array antennas. Phase array antennas consist of an array of antenna elements, the signals sent to which are weighted in both magnitude and phase. The combination of these weighted signals when radiated by the elements simultaneously form the antenna radiation pattern that can often be of complex shapes depending on the weights. In general, the antenna radiation pattern for an N-element array is represented by,

$$A(k) = a_0 \exp^{jkd_0} + a_1 \exp^{jkd_1} \ldots + a_{N-1} \exp^{jkd_{N-1}}$$

where a_n is a complex quantity corresponding to the magnitude and phase of the weight applied to the nth antenna element, $k = 2\pi$ and d_n represents the displacement of the element from a point of reference respectively. The applied weights help reinforce energy in a particular direction, thereby producing a high SNR (Signal to Noise Ratio) over an omni pattern in the desired direction contributing to a *direction/array* gain. Since phase array antennas available in practice cannot completely eliminate the energy radiated in undesired directions, they do result in some spill-over of energy in the unwanted directions, which are referred to as the side-lobes. As the main lobe is made more thin (focused), the array gain increases. However, it also increases the spill-over into side lobes. This trade-off is captured in the form of *front-side lobe ratio* of any directional antenna.

To realize a switched beam antenna, several such beam patterns can be generated with a phase array antenna such that they cover the entire azimuth (360°), and a specific beam pattern is dynamically chosen from the available set during operation. In this paper, we use Fidelity Comtech's Phocus Array [2] for our experiments. This antenna is a circular array of eight elements arranged in a regular octagon. The antenna is electronically steerable, i.e., a specific beam pattern out of the several precomputed beams can be chosen from software on the fly. Figure 1(a),(b,(c) show different patterns created with the Phocus array antenna, and the corresponding weights are shown in Figure 1(d). Patterns (a) and (b)—provided with the phase array by Fidelity Comtech—have a half-power beamwidth of 45° and a front to side lobe ratio of 8dB. Pattern (c), that we generated, has a half-power beamwidth of 60° and a higher front to side lobe ratio of 18dB. Note that with an N element antenna, the minimum main lobe width we can achieve is approximately 360°/N. Hence, for thinner beamwidths than 45°, we need greater than eight elements in the antenna.

3   Experimental Methodology

In this section, we describe our testbed and the methodology for studying the effectiveness of switched-beam antennas in improving spatial reuse and node localization. Our experiments evaluate the effects of different parameters such as beamwidth, front-to-side lobe ratio, node locations (line-of-sight or non-line-of-sight) and transmit power on the directionality of switched beam antennas.

Metric. The ideal beam pattern for a switched beam antenna is a single strong beam producing a high SNR (over omni pattern) in the direction of the receiver with no or negligible side-lobes in all the other directions. However, practical beam patterns do have considerable side-lobes. Further, multipath propagation indoors complicates the situation by resulting in reflected components of the main beam and the side-lobes. Thus, the three main components contributing to the directionality of a switched beam antenna for a given receiver are (a) "very few" beams with a large received signal strength S_S over the omni signal strength S_O ($S_S > S_O$), (b) "large" number of beams with a large reduction in received signal strength compared to omni ($S_S < S_O$), since this represents interference suppression in several directions, and (iii) the beam with the largest gain ($S_S > S_O$) coinciding with the geographic beam oriented towards the receiver.

Though there are multiple components to directionality, not all components may be required by applications and the specific components impacting applications varies with the nature of the application. For improving *spatial reuse*, we require the first two components of directionality to be satisfied, while it is not important the the strongest beam coincide with the main geographic beam. On the other hand, for a *localization* application, wherein it may be acceptable if the signal spills over in several beam directions, it is imperative that the third component be satisfied.

Setup. Our experimental setup is shown in Figure 2. The setup contains one AP connected to the Phocus array antenna, and 11 receiver nodes distributed in our office building in different office rooms and cubicles. Each receiver node is a small form-factor PC equipped with mini-PCI 802.11 a/b/g cards based on the Atheros 5212 chipset. The nodes run Linux kernel v2.4.26 and the MadWiFi driver [9] and their WLAN Radios connect to external OMNI antennas with a gain of 6dBi.

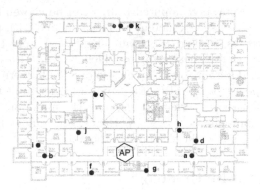

Fig. 2. Testbed Setup. Black dots indicate the locations of receiver nodes

For the Phocus array antenna, we use three beam pattern sets each containing eight patterns for eight directions to cover the entire circle (geographical area) around the AP. The patterns in each set are shifted by 45° from one another such that there is atleast one pattern that geographically covers each receiver. Pattern set 1 contains patterns like Figure 1(a), set 2 like Figure 1(b), and set 3 like Figure 1(c). We also generate an omni-directional pattern for comparison. All experiments are done on channel 6 in the 2.4Ghz spectrum at night to avoid disturbing and getting disturbed by regular office usage of the channel.

In all the experiments, the AP sends 128 byte UDP broadcast packets using the Click router package [8], and the receivers execute *tcpdump* in monitor mode. The AP utilizes the 802.11 pseudo-IBSS (Independent Basic Service Set) mode, in conjunction with monitor mode, which allows (a) all nodes to communicate directly and (b) the transmission of 802.11 broadcast frames at specified bit-rates from user-space. The AP chooses different directions in turn and transmits, and the receivers act as sensors by collecting data that helps us determine the directionality of the antenna.

4   Evaluation

In this section, we present several results that demonstrate in indoor environments (1) the degree of directionality obtained with a switched beam antenna, (2) the potential for spatial reuse, and (3) the accuracy of node localization.

Directionality. We first perform experiments to study the degree of directionality provided by the phase array antenna. In this set of experiments, the AP uses one of the three sets of beam patterns described in Section 3. For each set of beam patterns, the AP chooses a pattern in turn and broadcasts 1000 packets of size 128 bytes at 2 Mbps bitrate. From the received packets for each beam, we calculate the average RSSI on each receiver when the AP chooses the beam. We also repeat the experiment with an omni-directional pattern (OMNI). We then calculate the difference in RSSI between each beam and OMNI; if the

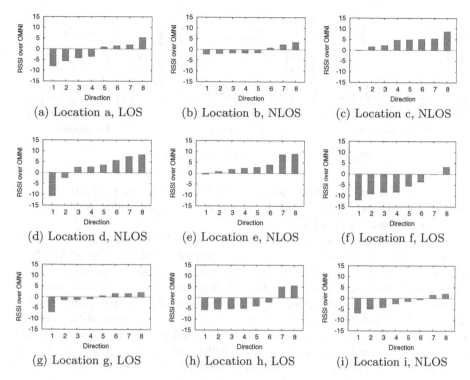

Fig. 3. RSSI over OMNI (in dB) at each receiver with different beams (directions)

difference is lower than 0 for beam j for a receiver, it means that the AP causes less interference on the receiver when it uses the beam j instead of OMNI.

Figure 3 shows our observations on different receivers for pattern set 3. We include the graphs for pattern sets 1 and 2 in a technical report [4] for brevity; the results look similar to set 3. As conventional wisdom says, we indeed find that there are several directions in which there is significant interference compared to OMNI 3(c,d,e). This observation is true even with beamwidths as low as 45° (with pattern sets 1 and 2) and front-side lobe ratio as high as 18dB (with set 3). Also, in a few cases, the strongest beam does not correspond to the main geographic beam 3(b,e,g). However, there are also many cases (a,b,f,g,h,i), and notably (a,f,h), where RSSI is significantly lower than OMNI in several directions. In a large number of cases, the strongest beam coincides with the geographic beam 3(a,c,d,f,h,i). With respect to the first two components of directionality, we find that the clients that indicate good directionality 3(a,f,h) are those that have a strong line-of-sight (LOS) component. However, this does not necessarily mean that all clients with a LOS component will show good directionality (e.g.3(g)) since it is possible for the multipath reflections to weaken the LOS component. With respect to clients with non-line-of-sight (NLOS) components, we find that they suffer in directionality with respect to the first two components. When it comes to strongest beam coinciding with the geographic

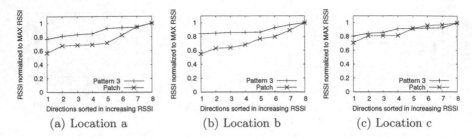

(a) Location a   (b) Location b   (c) Location c

Fig. 4. Directionality with a 17° patch antenna

beam, we find that this component of directionality is not dependent on the presence of a strong LOS component, with both LOS and NLOS clients showing good directionality 3(a,c,d,f,h,i).

These observations indicate that *the presence of multipath in indoor environments does not completely negate the directional benefits of switched beam antennas. Further, the availability of a strong LOS component does not indicate better directionality and vice versa. In fact, depending on the specific application considered and the specific components of directionality required by the application, both LOS and NLOS clients can potentially exploit directionality.* However, this would also require that the solutions designed for these applications take into account the implications of multipath reflections. To understand the importance of these implications, we further conducted experiments comparing the directionality of the 60° pattern produced by the phased arrays with that of a 17° pattern produced by a much bulkier patch antenna. We use a patch antenna here since an 8-element phase array antenna that we have can only generate a minimum beamwidth of 45°. The results presented in Figure 4 measure the normalized RSSI (sorted and normalized to maximum) as a function of different directions for three different clients (a), (b) and (c). The graphs demonstrate that thinner beam-widths do provide better directionality (by showing that the patch line is lower on the left than pattern 3).

Spatial Reuse. Ideally, determining the exact amount of spatial reuse possible is hard since it depends on several factors such as network topology, propagation characteristics in the indoor environment, MAC implementation, etc. Hence, we take an indirect approach to argue the potential for spatial reuse with switched-beam antennas. We observe that for spatial reuse, it is not just sufficient for the RSSI to be lower than OMNI, but should be lower by a considerable amount. Clearly, cases such as in Figure 3(a,f,h,i) show significant reduction in RSSI in multiple directions, thereby making spatial reuse possible. Just for illustration, we plot in Figure 5, the number of directions that are lower than OMNI by more than 3dB (every 3dB decrease represents halving the power, and hence interference). The graph shows that five locations have at least three directions (about 135 degrees) where interference is lower than OMNI by more than 3dB, thereby indicating chances of spatial reuse.

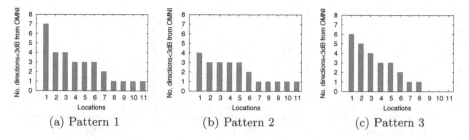

(a) Pattern 1   (b) Pattern 2   (c) Pattern 3

Fig. 5. Potential for Spatial Reuse with Switched-beam antennas

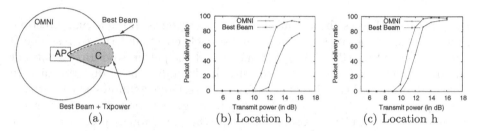

(a)   (b) Location b   (c) Location h

Fig. 6. Combining directionality and transmit power control for increased spatial reuse

Notice, however, that a directional antenna merely changes the area of interference compared to OMNI. To illustrate, Figure 6(a) shows a simplified picture of the area of interference with OMNI and directional beam patterns for an (AP,client) transmission. While the pattern "Best Beam" avoided interference in some regions that OMNI doesn't, "Best Beam" also causes interference in regions where OMNI doesn't due to the gain obtained by refocusing the transmission energy. Nevertheless, the AP can reduce the transmit power to the client for achieving the same performance as OMNI and meet the client's requirements. The effective beam looks like the shaded area in the figure that reduces the interference. Note that power control also reduces side lobes that cause interference to other transmissions. Figure 6(b) and (c) support the above proposal for combining transmit power control and directional transmission. In these graphs, we plot for two non-line-of-sight clients the MAC level packet delivery ratio obtained with changing transmit power, when using 54Mbps bitrate with OMNI and the best directional beam. The graphs clearly show that for the same packet delivery ratio, directional beam enables reducing transmit power over OMNI, thereby promoting higher spatial reuse.

Localization. Finally, we perform experiments to determine the effectiveness of the antenna in node localization. Most localization techniques assume that a receiver node is in the direction of the beam that produces the highest RSSI at the receiver, which might not be valid in reflection-rich environments. In this experiment, we measure the angle of deviation between the actual direction and the direction with highest RSSI for each of the receivers when the AP transmits with different beams from each pattern set. We also measure the difference in

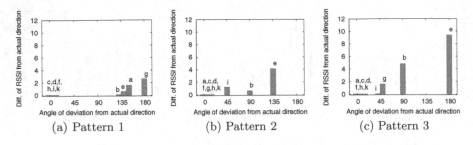

Fig. 7. Efficacy of localization

RSSI between the best beam and the beam that points to the actual direction. Figure 7 shows the angle of deviation for each client and the corresponding difference in RSSI (in dB). The graph shows that the patterns position significant number of clients correctly, satisfying the third component of directionality. In fact, some of these clients do not exhibit good interference suppression, but this is not required for localization. A few of the clients get positioned wrongly because of small difference in RSSI. Also, some clients that get wrongly positioned by one pattern get correctly positioned by another pattern. This observation suggests that in reflection-rich environments, just using the beam with highest RSSI may not be the right approach; a localization technique should choose from one of several best beams and even across different pattern sets for better accuracy. Developing such techniques is outside the scope of this paper.

Comparison with MIMO. Multiple-input multiple-ouput (MIMO) antenna technology is a popular alternative for indoor environments due to their ability to positively leverage multipath. However, for the specific applications of spatial reuse and node localization that we are interested in, MIMO antennas may still not serve the purpose. This is because (i) in the open-loop mode (no channel state information from receivers) MIMO can only help improve the link performance but cannot contribute to spatial reuse or address node localization, and (ii) in the closed-loop mode (having channel state information from receivers) they require significant feedback overhead and client modifications to enable adaptive beamforming that can address multipath. Nevertheless, a quantitative comparison to MIMO in terms of spatial reuse is an interesting topic for future work that we intend to take up as soon as such hardware and software become friendly for experimentation.

5   Conclusion

We study the directionality of a switched beam antenna in indoor environments in the context of spatial reuse and node localization. We make the following key observations. (1) Although reflections in indoor environments increase the interference in more directions than ideal, there can be several locations in an indoor environment where a directional beam indeed suppresses interference, thereby making spatial reuse possible. (2) For localization, while several clients get localized correctly, a few clients get wrongly positioned mainly because of a *small* difference in RSSI between the best beam and the beam in the physical direction of a client, thereby necessitating intelligent beam resolution mechanisms. (3) Finally, while thinner beamwidths do yield greater directionality, they do not completely eliminate the impact of reflections, which reinforces the importance of the above implications in indoor environments even with thin beams.

References

1. Hyperlink technologies, http://www.hyperlinktech.com
2. Phocus Array Antenna by Fidelity Comtech, http://www.fidelity-comtech.com/
3. Ruckus wireless smart antenna, http://www.ruckuswireless.com/
4. Blanco, M., Kokku, R., Ramachandran, K., Rangarajan, S., Sundaresan, K.: On the Effectiveness of Switched-beam Antennas in Indoor Environments, Technical report, NEC Laboratories America (2007), http://www.nec-labs.com/ravik/RESEARCH/dir-tech.pdf
5. Carey, J.M., Grunwald, D.: Enhancing wlan security with smart antennas: a physical layer response to information assurance. In: VTC (September 2004)
6. Casas, E., da Silva, M.T.C., Yin, H., Choi, Y.-S.: Beam diversity for indoor wlan systems. In: VTC (2003)
7. Choudhury, R.R., Yang, X., Ramanathan, R., Vaidya, N.H.: On designing mac protocols for wireless networks using directional antennas. IEEE Trans. Mobile Comput. 5(5), 477–491 (2006)
8. Kohler, E., Morris, R., Chen, B., Jannotti, J., Kaashoek, M.F.: The click modular router. ACM Trans. Comput. Syst. 18(3), 263–297 (2000)
9. MADWiFi. Multiband Atheros Driver for WiFi, http://madwifi.org
10. Navda, V., Subramanian, A.P., Dhanasekaran, K., Timm-Giel, A., Das, S.: Mobisteer: using steerable beam directional antenna for vehicular network access. In: MobiSys (2007)
11. Niculescu, D., Badrinath, R.: VOR base stations for indoor 802.11 positioning. In: Proc of MobiCom 2004, Philadelphia, PA (September 2004)
12. Park, J.-S., Nandan, A., Gerla, M., Lee, H.: Space-mac: enabling spatial reuse using mimo channel-aware mac. In: ICC (May 2005)
13. Raman, B., Chebrolu, K.: Design and evaluation of a new mac protocol for long-distance 802.11 mesh networks. MobiCom (2005)
14. Ramanathan, R., Redi, J., Santivanez, C., Wiggins, D., Polit, S.: Ad hoc networking with directional antennas: a complete solution. IEEE J. Sel. Areas Commun. 23(3), 496–596 (2005)
15. Sayrafian-Pour, K., Kaspar, D.: Source-assisted direction estimation inside buildings. In: INFOCOM (2006)
16. Vilzmann, R., Bettstetter, C., Medina, D., Hartmann, C.: Hop distances and flooding in wireless multihop networks with randomized beamforming. In: Proc of MSWiM 2005, Montreal, Quebec, Canada (October 2005)
17. Zhu, C., Nadeem, T., Agre, J.: Enhancing 802.11 wireless networks with directional antenna and multiple receivers. In: Proc. of MILCOM (2006)

On the Fidelity of 802.11 Packet Traces*

Aaron Schulman, Dave Levin, and Neil Spring

Department of Computer Science
University of Maryland, College Park
{schulman,dml,nspring}@cs.umd.edu

Abstract. Packet traces from 802.11 wireless networks are incomplete both fundamentally, because antennas do not pick up every transmission, and practically, because the hardware and software of collection may be under provisioned. One strategy toward improving the completeness of a trace of wireless network traffic is to deploy several monitors; these are likely to capture (and miss) different packets. Merging these traces into a single, coherent view requires inferring access point (AP) and client behavior; these inferences introduce errors.

In this paper, we present methods to evaluate the fidelity of merged and independent wireless network traces. We show that wireless traces contain sufficient information to measure their *completeness* and *clock accuracy*. Specifically, packet sequence numbers indicate when packets have been dropped, and AP beacon intervals help determine the accuracy of packet timestamps. We also show that trace completeness and clock accuracy can vary based on load. We apply these metrics to evaluate fidelity in two ways: (1) to visualize the completeness of different 802.11 traces, which we show with several traces available on CRAWDAD and (2) to estimate the uncertainty in the time measurements made by the individual monitors.

1 Introduction

Studying wireless networks "in the wild" gives researchers a more accurate view of 802.11 behavior than simulations alone. Researchers deploy monitors at hotspots such as cafes or conferences [10], or measure other deployed networks [1], to obtain traces of MAC and user behaviors. These traces provide realistic models of mobility [18, 11] and interference [1, 3] and many traces are readily available through sites such as CRAWDAD [7].

However, traces of real wireless networks have their own errors or assumptions. Indeed, capturing a high-quality wireless trace requires great care. Using too few monitors, placing them poorly, or using inadequate hardware can introduce missed or reordered packets and incorrect timestamps [16, 17, 10]. If multiple monitors are used, a *merging* algorithm combines the independent traces into a single view of the wireless network [10], but this process may order

* This work was supported by NSF-0643443 (CAREER). Dave Levin was supported in part by NSF Award CNS-0626964 and NSF ITR Award CNS-0426683.

M. Claypool and S. Uhlig (Eds.): PAM 2008, LNCS 4979, pp. 132–141, 2008.
© Springer-Verlag Berlin Heidelberg 2008

packets incorrectly. These potential errors mean that publicly available wireless traces vary greatly in quality (§5). Researchers must decide for themselves which wireless trace will provide them the most accurate, reproducible results.

We consider the problem of measuring the *fidelity* of wireless traces, which we decompose to their *completeness*—what fraction of the packets that could have been captured in fact were—and the *accuracy of their timestamps*. Our work is motivated by others' observations on how to use and improve the data that drives the networking community. As Paxson [12] notes, it is beneficial to identify how closely a measurement compares to reality before using it as experiment data. Haeberlen et al. also observe that researchers may fall into the trap of inappropriately generalizing their results if based on very specific or perhaps error-ridden data [8]. The difficult nature of capturing wireless traces further motivates a set of metrics and systematic means of measuring their quality.

We discuss how wireless trace fidelity can be measured by exploiting information in the trace (§3); external validation data is rarely available. We analyze a scoring method for wireless traces (§4). The percent of packets captured has been thought to be sufficient for quantifying a trace's fidelity, but we show that a richer description of fidelity is important and propose a way to visualize trace completeness that incorporates load (§5). We present several case studies from the CRAWDAD repository. We then study the accuracy of monitor and beacon timestamps, showing that clock accuracy is largely inversely proportionate to load and that clocks may need to be synchronized more frequently than at beacon intervals (§6). We conclude with lessons learned and directions for future work (§7). http://www.cs.umd.edu/projects/wifidelity holds our code and results.

2   Related Work

Because wireless traces are imperfect, many researchers have sought to improve trace fidelity. Yeo et al. [17, 16] and Rodrig et al. [14] discuss the steps they took to obtain high-fidelity traces, and use missing packets (§4) as a measure of fidelity. We focus on the relationship between trace quality and load on the monitor, and compare existing traces using our metrics.

Wit [10] attempts to refine existing traces by inferring and inserting missing packets. We believe traces that are as complete as possible at the time of capture are preferable, but that more complete traces will help the missing packet inference. Our tools are intended to help guide researchers toward capturing better traces and choosing the trace that best suits their needs.

Wireless traces are used for many reasons: to validate models of wireless behavior, study usage characteristics, and so on. Jigsaw [5, 4] uses wireless traces to measure and troubleshoot wireless networks. We emphasize that these pieces of work evaluate the *network*, and not the trace. We expect our work to complement these and other similar projects as pathologies in the input trace data could easily lead to false diagnosis by troubleshooting tools.

3   Self-evident Truths of Wireless Traces

Ideally, one could determine a trace's fidelity by comparing it to "truth": a perfect, complete trace of what was sent and when. In practice, only the trace itself is available. We show how the information in a wireless trace itself can be used to measure the trace's fidelity by detecting missed packets and measuring clock skew, and discuss the limitations of our methods.

3.1   Core Data in Wireless Traces

Traces vary in the information they include. Some traces have timestamps precise to nanoseconds, others only to milliseconds; not all traces record 802.11 acknowledgments; to maintain users' anonymity, few researchers release full payloads, and so on [15, 13]. The following data are available in all 802.11 CRAWDAD traces; we assume them as the *core* data that are likely to be available in future wireless traces:

1. All types of data packets.
2. All types of management packets including beacons, probe requests, and probe responses.
3. Full 802.11 header in all captured packets, including source and destination addresses (possibly anonymized), sequence number, retransmission bit, type, and subtype. Beacon packets also have timestamps applied by the AP.
4. Monitor's timestamp (set by the kernel or possibly the device).

3.2   Detecting Missed Packets

Monitors can fail to capture a packet because the monitor is overloaded, because there is interference and perhaps no stations receive the packet, because the signal is too weak at the monitor, and so on (Fig. 1). A common practice to reduce the number of missed packets is to place each monitor near an AP.

Most packet loss at the monitor can be inferred from 802.11 sequence numbers and the retransmission bit. When initially transmitted, each host (AP and client) assigns a packet a monotonically increasing sequence number from 0 to 4095 (or 2047 in some Cisco APs), and sets the retransmission bit to zero. One sign of missed packets is a gap in captured sequence numbers from a given host. Another sign of missed packets is a retransmitted packet without the corresponding first (non-re)transmission.

Missed retransmissions are more difficult to infer. Upon retransmission, the packet's sequence number remains unchanged, but the retransmission bit is set to one; future retransmissions of this packet are identical, which means that not all retransmissions can be inferred. If 802.11 acks and accurate timestamps are available, some of these retransmissions could be inferred. For instance, if a monitor captures an ack that is too late to correspond to any captured retransmission, we could infer that there must have been another retransmission. We do not consider this approach further, since not all traces contain acknowledgments.

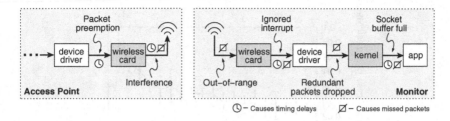

Fig. 1. Example sources of packet loss or timing errors in capturing wireless traces

3.3 Detecting Incorrect Timestamps

Monitors apply a timestamp to every packet in the kernel or possibly in the wireless device itself. The accuracy of these timestamps is vulnerable to delay at the AP and clock skew or clock drift at the monitor. Delay at monitors can come for many reasons, some of which we show in Fig. 1.

Beacon packets serve as a source of "truth" in that they allow us to synchronize the monitor's clock [10, 5]. However, this introduces its own sources of inaccuracy; timestamps in the beacon packets are subject to delay errors at the AP. Delay at the AP comes predominately in times of high load. When it is time to send a beacon packet, the AP creates the payload (including the timestamp), and attempts to send it. The timestamp in the beacon packets denotes when the packet was *created*, not necessarily when it was *sent*. Under high load, the packet may be stalled until the medium becomes free [2], increasing the difference between the packet's timestamp and when it was actually sent.

4 Scoring a Wireless Trace's Completeness

We propose a method to *score* wireless trace completeness. We value *completeness*—the fraction of packets captured—with the expectation that the more complete a trace is, the more useful it is. In the following section, we use our score along with traffic load to visualize completeness.

4.1 Estimating the Number of Missed Packets

Our scoring method is based on the number of missing packets from the wireless trace. This is an extension of what was introduced by Yeo et al. [16]. We define $\mathcal{P}_t$ to be the number of packets that should have appeared over time t.

$$\mathcal{P}_t \overset{\text{def}}{=} \sum_{nodes} \text{SeqNumChange}_t + \sum_{nodes} \text{Retransmissions}_t$$

The number of missing packets during time t, $\mathcal{M}_t$, is the number of packets that should have been captured minus the number of packets that were captured:

$$\mathcal{M}_t \overset{\text{def}}{=} \mathcal{P}_t - \sum_{nodes} \text{NumPacketsCaptured}_t$$

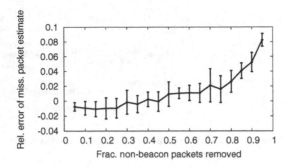

Fig. 2. Validation of our missing packets estimation. Starting with a high-quality trace (the Portland State University ug trace [13]), we remove non-beacon packets uniformly at random. Error bars represent 95% confidence intervals.

To evaluate the accuracy of this expression, we apply it to traces that we intentionally degrade. Starting with a high-quality trace (the Portland ug trace), we created progressively lower-quality traces by removing non-beacon packets uniformly at random and computed our score on these degraded traces (we expect monitors to capture most beacon packets: §5). We present the error of our missing packets estimation in Figure 2. Ideally, our method would detect all of these removed packets, but it is impossible to detect missing retransmission packets without 802.11 acknowledgments (§3). Even with a drastically degraded trace missing 95% of non-beacon packets, our score underestimates actual packet loss by only 10%. For more reasonable packet loss, our score has less than 5% error. These results indicate that *this method of detecting missing packets is accurate for both high- and low-quality traces.*

4.2   Score Definition

We define the *score* of a wireless trace's completeness during time t, $\mathcal{S}_t$, as the fraction of packets captured during time t: $\mathcal{S}_t \overset{\text{def}}{=} 1 - \frac{\mathcal{M}_t}{\mathcal{P}_t}$. Both APs and clients increment an independent sequence number for each unique packet transmitted. The technique used to reveal missing packets sent by an AP can do the same for clients. Unlike APs, clients do not transmit beacon packets at a regular interval. We must therefore be careful to keep track of how long it has been since the monitor last received a packet from a given client, so as to distinguish loss from, say, mobility. Our scoring method is subject to the same limitations as the missing packet estimation; the score cannot identify missing retransmissions.

5   Visualizing Wireless Trace Completeness

Trace completeness is an important component of fidelity. Rodrig et al. [14], for example, have used the percent of packets captured, similar to the score from §4, but we find a single number to be insufficient. This is in part because trace

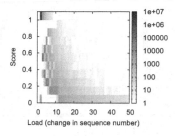 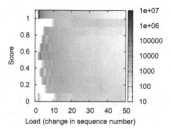

Fig. 3. Example T-Fi plots from the Sigcomm 2004 "chi" dataset, with scoring for only the AP (left), and scoring for APs and clients in a BSS (right)

quality can depend on load. A monitor may appear to capture a high percentage of packets, and one may be inclined to use that percentage to quantify the quality of a trace, but this number is misleading. For example, the Sigcomm 2004 trace "chi" contains 81% of AP data and management transmissions on channel 11. This percentage does not reveal that 37% of the packets collected were beacon packets sent when the AP was idle; not sending any other data or management packets. Excluding beacon packets sent during otherwise idle times, the monitor only saw 70% of the AP's transmissions.

5.1 T-Fi Plots

To overcome this problem, we visualize the score with a colormap, as shown in Figure 3. We refer to the colormaps as *T-Fi* or *Trace Fidelity* plots. The x-axis denotes the load from an epoch (beacon interval) in terms of the sequence number change during that epoch, and the y-axis denotes the score for that load. Color intensity denotes how often that (x, y)-pair occurred throughout the trace. The T-Fi plot displays these trace features:

1. The location on the y-axis shows completeness.
2. The width of the shaded region on the x-axis shows the range of load.
3. The intensity of the shaded region shows the frequency of load.

An ideal trace would have no missing packets and therefore a score of 1; in our visualization, this corresponds to a dark bar only at the top of the graph (the closest example of this is the Portland UG trace in Fig. 4).

Fig. 3 (left) shows how the single number problem can be overcome with a T-Fi plot. The darkest point on the plot is in the upper left hand corner. The upper left hand corner (sequence number change 1 and score 1) represents idle time beacon packets sent from an AP. The number of beacon intervals in this trace that fell in this region is 100 times larger than any other region in the plot. This would dominate a simple percentage, but is relegated to a small, clear region of the T-Fi plot. For load between 30 and 50, the trace scores no greater than 0.1, indicating low fidelity under high load. Indeed, Fig. 3 (left) shows a negative correlation of fidelity to load.

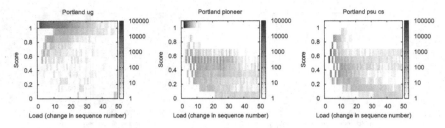

Fig. 4. Trace completeness visualization for Portland PDX traces [13]

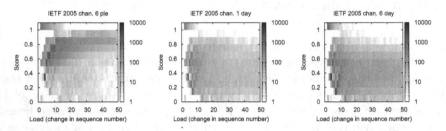

Fig. 5. Trace completeness visualization for IETF 2005 conference traces [9]

5.2   Case Studies

We analyzed the completeness of several traces obtained from CRAWDAD using the T-Fi visualization. We show two sets of traces: the Portland PDX VWave dataset and traces collected during the 2005 IETF meeting. Monitors from these traces may have captured unintended traffic from outside sources. The T-Fi plots shown in Figs. 4 and 5 are filtered to show only the BSS with the highest traffic.

Portland PDX. traces show how specialized 802.11 monitor equipment can improve trace quality. Phillips et al. [13] used a VeriWave WT20 commercial wireless monitor to capture their traces. VeriWave has a hardware radio interconnect to provide real time merging with 1 microsecond synchronization accuracy. UG has the best combination of high score and load. UG's T-Fi plot has a wide shaded region scoring 1 covering load values 1 to 40. This trace is close to complete and contains both high and low load epochs; Fig. 3 (left) represents a comparatively incomplete trace.

The pioneer trace (Fig. 4 center) was captured from an outdoor courtyard. Even with powerful monitor hardware, the monitor missed many packets in the pioneer trace. The trace contains a wide range of load values (1 to 50) but rarely scored above 0.5 in higher load epochs. Evidently, the pioneer trace is missing packets independently of the load. We believe the clients and AP captured by the trace were out of range or the monitor was receiving interfered signals. The psu-cs T-Fi plot (Fig. 4 right) has few dark-colored regions, indicating that there was low load on the network.

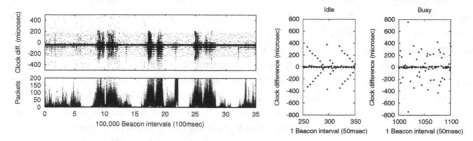

Fig. 6. Difference in monitor timestamps and beacon timestamps for the Sigcomm'04 "chi" trace (top left), with the load shown (bottom left). A controlled experiment with 50msec beacon intervals without load (middle) and with (right).

IETF 2005. traces exhibit high score variability under any given load. A load that scores consistently is represented in a T-Fi plot by a column that has only a few dark bars close together. This can be seen at sequence number change 40 on the T-Fi plot of "chan 6 ple" in Fig. 5. If the score varies greatly for a sequence number change the column will consist of similar colored bars; "chan 1 day" shows this behavior between sequence number changes 10 and 40.

The traces captured during the plenary sessions are of higher quality than the day sessions, showing the apparent effects of mobility on trace completeness. T-Fi plots of the day traces in Figure 5 do not score as highly as the plenary trace. For example, the plenary session traces score higher in high bandwidth epochs. We posit that the day traces scored lower in high bandwidth epochs because clients are mobile during the day. During the plenary sessions, the meeting participants were likely to be stationary more often than in the day traces.

6   Timestamp Accuracy

The accuracy of a trace's timestamps is important for many applications; merging algorithms [5, 10], for instance, use monitor and beacon timestamps to form a single, coherent view of the wireless network as viewed from potentially many monitors. A common assumption in these algorithms is that the difference between a monitor's timestamp—stamped in the kernel or the device itself—and the AP's timestamp—included in the beacon packet—is predictable and consistent on at least the order of beacon intervals (100msec).

We test this hypothesis by observing the difference between monitor timestamp and beacon timestamp over time throughout a trace. For the Sigcomm'04 trace (Fig. 6 left), we plot the clock difference (top) and the load in number of packets captured (bottom). The clock difference is not consistent from one beacon interval to the next, indicating that there is clock skew at the monitor and/or the AP. To see whether the clock difference was at least consistent *within* a given beacon interval, we collected our own trace using the MeshTest testbed [6] with a beacon interval of 50msec. When no clients are sending data (Fig. 6 middle), the clock difference does change *between* normal (100msec) beacon intervals, but in what appears, in this case at least, to be a predictable manner. However, when a client is sending (Fig. 6 right), the clock changes are not predictable, again indicating a correlation of clock difference with load.

These results show that *the common assumption underlying known merging algorithms is false.* The question remains whether this is sufficient to cause a mis-ordering of packets. Though we have observed mis-orderings from Wit [10], it is unclear whether this is due to an algorithmic error or simply a bug in Wit. Nonetheless, we propose as a sanity check that merging algorithms ensure proper sequence number order (not necessarily strictly increasing: §7).

7   Discussion

We considered the problem of quantifying wireless trace fidelity and evaluated a scoring method, proposed the T-Fi visualization, and presented an analysis of clock accuracy in wireless traces. Wireless trace fidelity applies when choosing, improving, or inferring gaps in wireless traces.

Choosing a trace. Researchers will choose traces from a repository like CRAW-DAD based primarily on the type of data in the trace, for example mobility or traffic type. However, we expect fidelity to decide which trace—or subset of the trace—to use.

Improving traces. Measuring trace fidelity need not be strictly a post-mortem analysis; rather, researchers ought to measure the fidelity of their measurements *during* their measurement, so that they may, for example, move their monitors. An interesting and important area of future work is to develop tools to aid in the active capture of wireless traces, so that researchers can ensure high-fidelity traces in unique hotspots such as a conference.

We conclude with lessons we learned about merging and processing wireless traces in the process of working with as many traces as we could collect.

Update tools in accordance with new specs. Tools to measure the fidelity of wireless traces must be updated frequently, as new 802.11 specs are deployed. The 802.11e QoS amendment introduced a new sequence number space for QoS in mid-2006. This did not turn up in our initial testing on the Sigcomm'04 trace, but did in the Portland traces (late 2006), and we had to adjust our tool accordingly.

Account for vendor-specific behavior. Some vendors introduce behavior not specified in 802.11, and this may make the trace appear to be of lower fidelity. We observed that the Cisco access point in the Sigcomm'04 trace assigned sequence numbers to broadcast and multicast packets, then transmitted the packets after others were sent, causing some sequence numbers to appear out of order. To account for this, we allowed these packets to appear out of order in sequence number.

Acknowledgements. We thank Justin McCann and the anonymous reviewers for their helpful comments, Brenton Walker and Charles Clancy for allowing us to use the MeshTest testbed, and Ratul Mahajan for supporting Wit.

References

1. Aguayo, D., Bicket, J., Biswas, S., Judd, G., Morris, R.: Link-level measurements from an 802.11b mesh network. In: SIGCOMM (2004)
2. ANSI/IEEE. Std 802.11 (1999)
3. Biswas, S., Morris, R.: Opportunistic routing in multi-hop wireless networks. In: SIGCOMM (2005)
4. Cheng, Y.-C., Afanasyev, M., Verkaik, P., Benkö, P., Chiang, J., Snoeren, A.C., Savage, S., Voelker, G.M.: Automating cross-layer diagnosis of enterprise wireless networks. In: SIGCOMM (2007)
5. Cheng, Y.-C., Bellardo, J., Benkö, P., Chiang, J., Snoeren, A.C., Voelker, G.M., Savage, S.: Jigsaw: Solving the puzzle of enterprise 802.11 analysis. In: SIGCOMM (2006)
6. Clancy, T., Walker, B.: MeshTest: Laboratory-based wireless testbed for large topologies. In: TridentCom (2007)
7. CRAWDAD Website, http://crawdad.cs.dartmouth.edu/
8. Haeberlen, A., Mislove, A., Post, A., Druschel, P.: Fallacies in evaluating decentralized systems. In: IPTPS (2006)
9. Jardosh, A., Ramachandran, K.N., Almeroth, K.C., Belding, E.: CRAWDAD data set ucsb/ietf (v. 2005-10-19) (October 2005), Downloaded from http://crawdad.cs.dartmouth.edu/ucsb/ietf2005
10. Mahajan, R., Rodrig, M., Wetherall, D., Zahorjan, J.: Analyzing the MAC-level behavior of wireless networks in the wild. In: SIGCOMM (2006)
11. Navidi, W., Camp, T.: Stationary distributions for random waypoint models. IEEE Transactions on Mobile Computing 3(1) (2004)
12. Paxson, V.: Strategies for sound Internet measurement. In: IMC (2004)
13. Phillips, C., Singh, S.: CRAWDAD data set pdx/vwave (v. 2007-08-13) (August 2007), Downloaded from http://crawdad.cs.dartmouth.edu/pdx/vwave
14. Rodrig, M., Reis, C., Mahajan, R., Wetherall, D., Zahorjan, J.: Measurement-based characterization of 802.11 in a hotspot setting. In: E-WIND (2005)
15. Rodrig, M., Reis, C., Mahajan, R., Wetherall, D., Zahorjan, J., Lazowska, E.: CRAWDAD data set In: uw/sigcomm2004 (v. 2006-10-17) (October 2006), Downloaded from http://crawdad.cs.dartmouth.edu/uw/sigcomm2004
16. Yeo, J., Banerjee, S., Agrawala, A.: Measuring traffic on the wireless medium: Experience and pitfalls. Technical report, CS-TR 4421, University of Maryland, College Park (December 2002) http://hdl.handle.net/1903/124
17. Yeo, J., Youssef, M., Agrawala, A.: A framework for wireless LAN monitoring and its applications. In: WiSE (2004)
18. Yoon, J., Liu, M., Noble, B.: Random waypoint considered harmful. In: INFOCOM (2003)

Refocusing in 802.11 Wireless Measurement

Udayan Deshpande[1], Chris McDonald[2], and David Kotz[1]

[1] Institute for Security Technology Studies,
Department of Computer Science, Dartmouth College
[2] School of Computer Science and Software Engineering,
The University of Western Australia

Abstract. The edge of the Internet is increasingly wireless. To understand the Internet, one must understand the edge, and yet the measurement of wireless networks poses many new challenges. IEEE 802.11 networks support multiple wireless channels and any monitoring technique involves capturing traffic on each of these channels to gather a representative sample of frames from the network. We call this procedure *channel sampling*, in which each sniffer visits each channel periodically, resulting in a sample of the traffic on each of the channels.

This sampling approach may be sufficient, for example, for a system administrator or anomaly detection module to observe some unusual behavior in the network. Once an anomaly is detected, however, the administrator may require a more extensive traffic sample, or need to identify the location of an offending device.

We propose a method to allow measurement applications to dynamically modify the sampling strategy, *refocusing* the monitoring system to pay more attention to certain types of traffic than others. In this paper we show that refocusing is a necessary and promising new technique for wireless measurement.

1 Introduction

The new edge of the Internet is wireless. At Dartmouth College, all undergraduate students own wireless laptops, and take advantage of ubiquitous 802.11 coverage on campus. Most large enterprises provide wireless coverage for employee use. Many cities are deploying or considering large-scale municipal wireless-access networks. Although these networks were originally intended as a convenience, an overlay for the wired network, for many users and in many places the wireless network is of fundamental importance. To understand the edge of the Internet, we need effective wireless measurement.

Consequently, there are many motivations for monitoring wireless networks, including network management, security, and research. We consider situations where the network of wireless access points (APs) is augmented with interspersed wireless air monitors (AMs). These dedicated sniffers can provide real-time capture of wireless traffic and measurement of MAC-layer conditions for network analysis and management [6], and intrusion-detection systems (IDS) can analyze live streams of traffic from these AMs to monitor the network for attacks [2,9].

M. Claypool and S. Uhlig (Eds.): PAM 2008, LNCS 4979, pp. 142–151, 2008.

Wireless (802.11) networks allow traffic on multiple parallel channels, and yet all practical monitoring systems can listen to only one or two channels at a time. This approach is limited because there may be a need to monitor all the channels (e.g. to locate the presence of ad-hoc networks or rogue APs.) If there is only one channel to be monitored, the radio can simply monitor that channel continuously [6]. If no specific channel is desired, most scanning systems simply capture traffic on all channels with a predefined time spent on each channel [11]. Earlier work [2,6] acknowledges the need for smart channel-sampling strategies in security and management applications.

Our earlier work [9] demonstrates how to improve the capture by dynamically scheduling AMs to spend more time on channels with higher frame rates. In this work we extend our sampling philosophy by demonstrating a technique and framework that allows external applications— such as an administrator's console, or an IDS- to dynamically instruct the AMs to put more effort into capturing traffic that meets a given condition.

We describe the traffic trace that a monitoring application requires at any time by its *focus*. An application can, of course, filter the stream of captured frames to suit its interests, but we want to allow the application to *refocus* the measurement system to skew its ongoing traffic capture towards this new focus, capturing more of the desired frames. We recognize that many important scenarios require the capture of a baseline sample, suitable for basic monitoring by multiple applications, and simultaneously a more focused sample(s) required by one or more applications.

For example, an application may be content with a traffic trace that consists of equal samples from each of the channels being monitored in the network. After observing some event, it may wish to *refocus* most of the sampling effort on the channels where a specific MAC address was observed. This application could be a WLAN intrusion-detection system, an application that displays locations of 802.11 devices around an office, or a system that monitors the quality of voice-over-wireless calls.

We claim that dynamic refocusing helps the wireless-network measurement system be more responsive to the needs of the subscribers of measured data. We describe a method and a tool that enables refocusing.

2   Related Work

We draw on related work in wireless measurement and 802.11 security. Few large-scale 802.11 measurement studies have attempted to capture wireless frames from the air. Although a few papers characterize traffic at meetings and conferences [10,13], none consider channel sampling or refocusing. To our knowledge, no commercial products provide refocusing, although some do allow channel sampling; for example, Aruba Networks [1], and Kismet [11]. Our own earlier work [9] focused on the challenge of sampling traffic from many channels, and merging frames from many AMs; in this paper we look at the problem of refocusing through a large-scale experimental deployment. We compare one of the strategies used in the previous paper with our refocusing mechanism.

Security in 802.11 remains a challenge, because there are many vulnerabilities in the protocol and its implementations [2,4,14, for example]. We expect refocusing to help in capturing more information about an ongoing attack that is first detected during baseline sampling of a network.

A few recent papers describe offline tools to capture and merge wireless frames from multiple AMs located around a building [8,12,15]. These papers concentrate on methods for synchronizing traces collected across multiple AMs into a single chronological trace, inferring missing frames, reconstructing transport-layer flows, and detecting performance artifacts and network inefficiencies. Most of these tools work only on offline traces. One, *Jigsaw* [8], requires four radios per location, clearly a more expensive solution. When few AMs are available, each radio must sample many channels, and our system of refocusing helps to gather the most relevant information with limited resources. In Jigsaw [8], the authors place 39 monitoring "pods" around the building with four radios each. Each radio (AM) monitors a separate channel (Channels 1, 6, 11 and another "center" frequency). In their coverage experiments, their clients associate with APs and transfer data using scp. They report that their sniffers capture about 90% of all the scp frames sent to and from the clients. This experiment assumes that only traffic on the same channels as the APs that can be observed by *both* the AM *and* the client, or that can be observed by *both* the AM *and* the AP, needs to be monitored. There is no experiment in the paper that measures the coverage in the scenario where only the AP or the client is in the range of a transmitting radio but not the nearby AM. Due to the static allocation of channels to AMs, if there is an AM in range, it may be on a different channel. This case is, of course, important in a security scenario. With the increasing numbers of channels available for transmission in 802.11 networks, simply increasing the number of radios in a "pod" cannot be the answer. It is clear, therefore, that channel sampling is the only practical technique to cover a large monitoring area. The claim made in the Jigsaw paper [7] that monitoring platforms from DAIR [2] and Jigsaw provide "the ability to observe every link-layer network transmission across location, frequency and time" is overly optimistic.

The DAIR system [2] uses USB NICs to turn an enterprise's desktop computers into AMs, and could benefit from our sampling techniques for collection of traffic from production networks. The newer DAIR-based network management system [6] simply assigns the USB NICs to the channels of the nearby access points, missing important security-related traffic on non-production channels.

3   Dingo: A Coordinated Sniffer

We developed a set of software components, named *dingo*,[1] that collectively enable a variety of packet sampling policies to be defined and controlled, and

---

[1] A dingo is an Australian native dog renowned for its ability to track prey in bleak conditions.

their effects monitored. dingo comprises two main components: *amsniffer*, which runs on each AM device, and *amcontroller*, which runs on a more powerful central Linux server. dingo also employs an additional software component, a *merger* developed as part of earlier work, and described below. Figure 1 shows the principal components of this software and the communication paths between them.

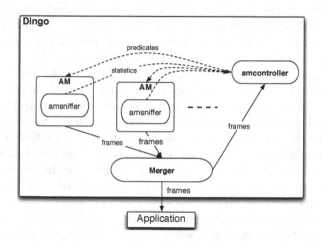

Fig. 1. The Sampling Architecture

The *amsniffer* component runs on each AM; multi-radio AMs can run an instance for each radio. (In our experiments we found that it is more effective to invoke two instances of amsniffer, each listening on a different interface, than it is for a single process to monitor two interfaces in an interleaved manner.) Command-line options to amsniffer indicate which wireless interface should be employed, the default sniffing policy to be followed, and the destination for captured frames.

The amsniffer captures features from each frame header and transmits it over a wired Ethernet infrastructure to the merger using UDP/IP. The role of the merger is to interleave the AMs' streams of frames into a chronologically consistent ordering, and to remove frames captured in duplicate by multiple AMs. For duplicates, the output record includes a list of the receiving AMs and signal strength. The merger's output is forwarded to subscribing applications and to our amcontroller.

The role of the amcontroller is to determine *scheduling policies* and to disseminate them to the AMs. Policies specify a sequence of channel numbers, and the duration for which the interface should listen on each channel. A typical scheduling *cycle* will involve visiting each channel, collecting a variety of statistics about the traffic observed on each channel. Each instance of amsniffer executes its current scheduling policy for a requested number of cycles or until directed by amcontroller to execute a new policy, either an existing pre-stored policy or one computed by the amcontroller. We found our devices can experience a significant delay when changing from one channel to another, and that this delay is minimized by visiting requested channels in ascending order (approx. 30ms when ascending, 300ms when descending), so we limit all schedules to this order, descending only once at the end of the cycle.

Notice that our approach does not require specific policies to be "hard-wired" into amsniffer. Each amsniffer may receive a distinct scheduling policy, perhaps determined from the type and extent of traffic recently sampled by that amsniffer and its neighbors, or to consistently monitor traffic in a particular geographical region. The ability to remotely program the AMs provides the greatest opportunity to experiment with new sampling strategies.

While sampling traffic, each amsniffer maintains a small number of counters, including the number of frames captured on each channel, the total length of those frames, and the number of frames matching one or more Boolean *predicates* provided by the amcontroller. At the end of each cycle, amsniffer sends its counters to the amcontroller for consideration in future scheduling decisions. The range of policies described in our earlier work [9] are based on these simple counts gathered at the AMs. For example, a policy employing *proportional sampling* spends time on each channel proportional to the recently observed frame rate on that channel.

The predicates are written in a small language, similar to C's expressions. The language supports all precedence levels, equality and relational operators, and data types including integer, Boolean, string, and MAC address. About 30 keywords in the language correspond to the attributes of each captured frame and the wireless environment in which it was captured. Our predicates provide access to the 802.11 header attributes and a few PHY-layer attributes, and are analogous to the expressions supported by the popular `tcpdump` utility and Berkeley packet filter. For example, predicates may determine whether a captured frame was a control, management, or data frame, may examine the source, destination, and BSSID MAC addresses of frames, examine a frame's length, payload length, the channel on which it arrived, or its relative signal strength.

To support refocusing, dingo's amcontroller uses the predicate counters in a modified form of proportional sampling, scheduling each amsniffer to spend time on each channel in proportion to the number of frames matching the predicate. In this manner, amsniffers focus on the traffic of interest, while still devoting a small amount of time on other channels to determine if the traffic pattern is observed there. For example, the predicate `"src == 00:16:cb:b7:18:82"` could be used to focus on traffic from a stolen laptop's wireless interface. Any amsniffers capturing frames matching this predicate will be instructed by the amcontroller to devote more sampling time to the channels recently carrying that traffic. AMs not capturing traffic from this laptop will continue to follow a default sampling policy. If the laptop associates with a different access point using another channel, or moves within range of different AMs, the shorter time spent on other channels will facilitate those AMs to focus on the laptop. A short cycle time, typically 1 or 2 seconds, enables each amsniffer to quickly identify and focus on required traffic patterns. Again, this ability to remotely program the AMs with a wide variety of predicates facilitates experimentation.

4   Applications of Refocusing

We believe that refocusing has many applications in wireless research, security, and network management. Any application that requires more than cursory scanning of the traffic in the wireless medium will sometimes desire an increased focus on some subset of the traffic, and yet other applications will simultaneously need a baseline broad sampling.

Filtering alone may not achieve the results needed by the application, because the necessary frames may not be available. Therefore, there is a need for online dynamic refocusing of the monitoring hardware.

We consider three classes of application.

Localization. If a WIFI device needs to be geographically localized, the refocusing system can focus more attention on it by capturing more frames to and from it. Refocusing may aid in better localizing the laptop, by capturing more frames from as many different perspectives (AMs) as possible. We can capture more samples in less time, increasing the accuracy or reducing latency for estimating the location of the laptop using any of the state-of-the-art methods. We describe one such experiment in Section 5.2.

VOIP-quality measurement. Consider an enterprise network manager who wishes to monitor the quality of Voice-over-IP calls. If there are known VoIP clients using the Wi-Fi network, we can focus on those MAC addresses and thus monitor the relevant channels, more closely. Alternately, we could focus on channels with observed VoIP activity (by recognizing the use of particular protocols) or through a higher-level metric like the jitter, per-frame delay in the VoIP calls, or the observed congestion in a channel. For example, the predicate may take the form "jitter $>= x$ ms". Such high-level predicates cannot yet be matched in dingo. This capability is part of our future work.

Security monitoring. For example, we can refocus on channels that carry an excessive number of deauthentication messages, or on MAC addresses that are known to have been recently spoofed. In the future, using our techniques, we can focus on channels where new clients appear, then study their packets to discern whether they seem especially vulnerable to attack. The system can fingerprint new clients to determine if they are employing drivers, cards, or operating systems with known vulnerabilities [5]. If indeed they are vulnerable, we can refocus our sampling to more closely monitor them.

5   Results

We set out to investigate whether refocusing can be a valuable tool in wireless measurement systems. In this short paper, we do not have the space for a complete evaluation, but we seek to demonstrate the potential value of this approach.

5.1   Improved Volume of Capture

In our CS department, we deployed 19 Aruba AP70 AMs throughout the three floors of the building. The building also has 20 802.11a/b/g access points. The AP70 has a MIPS IDT32434 CPU running at 266MHz, 32MB DRAM, two Atheros AR5212 802.11a/b/g NICs (network interface controllers), two Ethernet NICs and one USB port. We installed OpenWRT Linux (Kamikaze branch, r5494) and Madwifi (v0.9.2) on each, and a copy of amsniffer on each. In this experiment, we only used one of the two wireless NICs on each AM.

We performed two experiments in which a laptop transmitted 10 UDP frames per second to the non-existent MAC address 22:22:22:22:22:22 on a channel randomly selected from the 11 802.11b channels. The laptop changed channels every 10 seconds. In each experiment, the laptop was carried around a fixed path in the CS department building for a period of 10 minutes.

In the first experiment our AMs used the traditional equal-time sampling strategy in which the AMs spend equal time on all the channels. In the second experiment, we refocused the AMs to spend more time on the channels that were observed to capture more frames from the experimental laptop, using the predicate "dst == 22:22:22:22:22:22".

Figure 2 plots the number of frames that matched the predicate, as seen in the output of the AMs in both cases. We can see that every AM consistently captured more frames from our mobile laptop when we ran the refocusing strategy than when we ran the equal-time strategy.

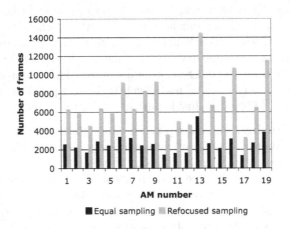

Fig. 2. Number of frames captured that matched predicate

In Figure 3 we present the number of frames that *did not* match the predicate. Although the refocused strategy captured fewer such frames than the equal-time strategy, it still provided a flow of such baseline traffic sufficient for use by other subscribers. That is, the refocusing requested by one application does not preclude ongoing monitoring by background activities, at least in this case.

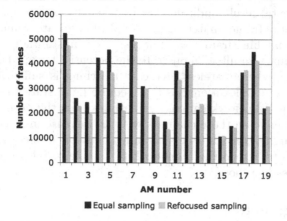

Fig. 3. Number of frames captured that did not match predicate

5.2 Localization Experiment

Our hypothesis is that refocusing will allow an application to more accurately, and more quickly, determine the location of a given wireless client. We chose a technique, the Nearest Neighbor in Signal Space (NNSS) method, described by Bahl et al. [3]. This localization algorithm uses observed signal strengths of frames heard by clients from APs. We used the dual of this algorithm and constructed the signal space by using the RSSI of captured frames from the client at AMs to populate our signal space.

Firstly, we calibrated the corridor of the third floor of our building. We measured the signal strength at every AM from the frames of a client transmitting 50 frames at every five feet along the corridor. In this phase, we configured all of the AMs to capture traffic on channel 1, and configured the client to transmit on channel 1. In the second phase, we configured the AMs to sample equally on every channel, and we captured a trace of the client transmitting 10 frames at every 10 feet along the corridor. Finally, we configured the AMs to refocus on the MAC address of our client and captured a trace of the transmissions of the client at the same locations as in the second case. With our refocusing mechanism we observe localizations that are, on average, 1.95 feet more accurate than without refocusing.

6 Discussion

Given that the Internet edge is increasingly wireless, and the increasing number of channels in Wi-Fi (802.11n includes far more channels than 802.11b, for example), any researcher or network manager seeking to measure their network traffic must find efficient mechanisms to sample the network traffic. We propose a mechanism for applications to dynamically refocus the attention of the wireless-measurement infrastructure to capture more of a desired kind of network traffic.

Our experimental results indicate that refocusing was successful in capturing a greater number of frames matching the supplied predicate. Simultaneously, the capture of non-matching frames was not degraded substantially.

These preliminary results demonstrate the potential for our framework to be used in scenarios where stateless predicate matching is sufficient. There may, however, be more complex scenarios that require more state to be maintained. For example, our framework cannot currently express the desire to refocus on *newly arrived* clients, on those channels with an *increase* in some metric, or on channels with high jitter (inter-arrival times between frames) in a voice flow. Our next step is to extend the framework to be able to refocus on the basis of temporal changes like those.

Acknowledgments

We gratefully acknowledge the input and support of colleagues on the MAP team, and Dartmouth's network administrators, including Wayne Cripps and Tim Tregubov. The staff at Aruba Networks has been invaluable, including Nick DePetrillo and Mike Baker.

This research program is a part of the Institute for Security Technology Studies, supported under award number NBCH2050002 from the U.S. Department of Homeland Security, Science and Technology Directorate. Points of view in this document are those of the authors and do not necessarily represent the official position of the U.S. Department of Homeland Security or the Science and Technology Directorate. This project was also supported by the Cisco Systems University Research Program, the Center for Mobile Computing at Dartmouth College, and NSF Infrastructure Award EIA-9802068.

References

1. Aruba Networks Air Monitors,
   http://www.arubanetworks.com/technology/air-monitors/
2. Bahl, P., Chandra, R., Padhye, J., Ravindranath, L., Singh, M., Wolman, A., Zill, B.: Enhancing the security of corporate Wi-Fi networks using DAIR. In: Proceedings of MobiSys 2006, Uppsala, Sweden, June 2006, pp. 1–14 (2006)
3. Bahl, P., Padmanabhan, V.N.: RADAR: An in-building RF-based user location and tracking system. In: Proceedings of InfoCom 2006, pp. 775–784 (2000)
4. Bellardo, J., Savage, S.: 802.11 denial-of-service attacks: Real vulnerabilities and practical solutions. In: Proceedings of the Twelfth USENIX Security Symposium, Washington, DC, USA, August 2003, (USENIX Association), pp. 15–28 (2003)
5. Bratus, S., Cornelius, C., Kotz, D., Peebles, D.: Active behavioral fingerprinting of wireless devices. In: Proceedings of the First ACM Conference on Wireless Network Security (WiSec), March 2008, ACM Press, New York (accepted for publication, 2008)
6. Chandra, R., Padhye, J., Wolman, A., Zill, B.: A location-based management system for enterprise wireless LANs. In: Proceedings of the 4th USENIX Symposium on Networked Systems Design and Implementation (NSDI 2007), Cambridge, MA, USA (2007)

7. Cheng, Y.-C., Afanasyev, M., Verkaik, P., Benkö, P., Chiang, J., Snoeren, A.C., Savage, S., Voelker, G.M.: Automating cross-layer diagnosis of enterprise wireless networks. SIGCOMM Comput. Commun. Rev. 37(4), 25–36 (2007)
8. Cheng, Y.-C., Bellaro, J., Benko, P., Snoeren, A.C., Voelker, G.M., Savage, S.: Jigsaw: Solving the puzzle of enterprise 802.11 analysis. In: Proceedings of SIGCOMM 2006, Pisa, Italy, September 2006, pp. 39–50 (2006)
9. Deshpande, U., Henderson, T., Kotz, D.: Channel sampling strategies for monitoring wireless networks. In: Proceedings of the Second Workshop on Wireless Network Measurements, USA, April 2006, IEEE Computer Society Press, Boston (2006)
10. Jardosh, A.P., Ramachandran, K.N., Almeroth, K.C., Belding-Royer, E.M.: Understanding congestion in IEEE 802.11b wireless networks. In: Proceedings of the 2005 Internet Measurement Conference, Berkeley, CA, USA, October 2005, pp. 279–292 (2005)
11. Kismet wireless sniffer, http://www.kismetwireless.net
12. Mahajan, R., Rodrig, M., Wetherall, D., Zahorjan, J.: Analyzing the MAC-level behavior of wireless networks in the wild. In: Proceedings of SIGCOMM 2006, Pisa, Italy, September 2006, pp. 75–86 (2006)
13. Rodrig, M., Reis, C., Mahajan, R., Wetherall, D., Zahorian, J.: Measurement-based characterization of 802.11 in a hotspot setting. In: Proceedings of the ACM SIGCOMM 2005 Workshop on Experimental Approaches to Wireless Network Design and Analysis (E-WIND-2005), Philadelphia, PA, USA (August 2005)
14. Wireless vulnerabilities & exploits database, http://wirelessve.org
15. Yeo, J., Youssef, M., Henderson, T., Agrawala, A.: An accurate technique for measuring the wireless side of wireless networks. In: Proceedings of the International Workshop on Wireless Traffic Measurements and Modeling, Seattle, WA, USA, June 2005, pp. 13–18 (2005)

Pathdiag: Automated TCP Diagnosis*

Matt Mathis[1], John Heffner[1], Peter O'Neil[2,3], and Pete Siemsen[2]

[1] Pittsburgh Supercomputing Center
[2] National Center for Atmospheric Research
[3] Mid-Atlantic Crossroads

Abstract. This paper describes a tool to diagnose network performance problems commonly affecting TCP-based applications. The tool, *pathdiag*, runs under a web server framework to provide non-expert network users with one-click diagnostic testing, tuning support and repair instructions. It diagnoses many causes of poor network performance using Web100 statistics and TCP performance models to overcome the lack of otherwise identifiable symptoms.

1 Introduction

By design, the TCP/IP hourglass [4] hides the details of the network and the application from each other. This property is critical to the ongoing evolution of the Internet because it permits applications and the underlying network infrastructure to evolve independently. However, it also obscures all network flaws. Since TCP silently compensates for flaws, for example by retransmitting lost data, the only symptom of most problems is reduced performance. This "symptom hiding" property was the motivation behind the Web100 project [17], which developed the TCP extended statistics MIB [16] to expose TCP protocol events that are normally hidden from the application. A MIB is a formal specification of a set of management variables that can be accessed by SNMP or other lower overhead mechanisms. Experimental prototypes of the MIB have been implemented in a number of operating systems, including Linux [17] and Microsoft Windows Vista [23].

Diagnostic efforts are further complicated by another property of TCP: the symptoms of most flaws scale by the flow's round-trip time (RTT). Note that for window-based protocols, performance models generally have an RTT term in the denominator. For example, insufficient TCP buffer space in either the sender or receiver, or background (non-congested) packet loss all cause TCP to have a constant average window size and performance that is inversely proportional to the RTT.

This poorly understood property leads to faulty reasoning about diagnostic results. A simple throughput test on a short local section of a path with minor flaws is likely to yield good results. The same test run over a longer path containing the same local flaws is likely to yield poor results. The naïve conclusion

---

* This work was supported by the National Science Foundation, Grant ANI–0334061.

M. Claypool and S. Uhlig (Eds.): PAM 2008, LNCS 4979, pp. 152–161, 2008.

would be that the local section is flawless, and the problem must be present in the longer path section. This "symptom scaling" property of TCP leads to incorrect inductive reasoning about flaws, and significantly contributes to the difficulty of solving end-to-end Internet performance problems.

This paper describes a tool, *pathdiag*, that uses TCP performance modeling to extrapolate the impact of local host and network flaws on applications running over long paths. The tool analyzes a number of key metrics of the local host and path and uses TCP performance models to determine thresholds for these metrics based on the stated application performance goals. *Pathdiag* reliably detects flaws that have no user-noticeable symptoms over a short path. It reports the problems and suggests remedies.

1.1 Motivation

Network performance has increased by an order of magnitude roughly every four years over the last two decades. Networking experts are usually quick to demonstrate the full data rate on each new network technology [11]. However, typical users experience data rates much lower than those seen by experts, and the gap is widening.

Internet2 has measured the performance of TCP bulk flows over their backbone since the beginning of 2002 [12]. As of August 2007, the median performance across their 10 Gb/s network was only about 3.4 Mb/s. Historical data shows that this rate has taken six years to double.

A small number of flows get very good performance. About 0.1% are faster than 100 Mb/s, and of those about half are close to 1 Gb/s. Since the backbone carries a significant number of very high-rate, long-distance flows, we know that it has to be free from flaws that would otherwise affect these sensitive flows.

The design goal of *pathdiag* is to help non-expert users attain better performance by easily and accurately diagnosing common flaws. These flaws are generally near the edge of the network where debugging efforts are subject to faulty inductive reasoning due to symptom scaling.

2 The *Pathdiag* Tool

Suppose a user tries to get good performance from an application that relies on bulk TCP data transfers from a remote server, as shown in Figure 1. The user's application client C, needs data from the application server S across a long network path that includes both a short local section and a long-haul backbone. The local section is assumed to have an RTT that is no more than a few milliseconds. The long-haul backbone can be any length, transcontinental (100 ms RTT) or even global (300 ms RTT).

The user can test the local section of the path and the client configuration by visiting a *pathdiag* server, PS, with a java-enabled browser. Ideally, PS would be located near the connection between the local network and the backbone. The *pathdiag* server tests the local path and client configuration and generates a report in the form of a new web page, displayed by the user's browser.

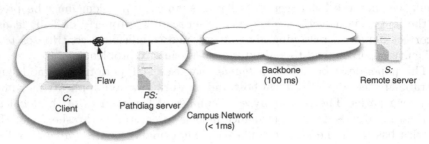

Fig. 1. Canonical *pathdiag* setup

Pathdiag estimates whether the local client and local path is sufficient to meet the target[1] data rate if the backbone were replaced by an ideal network with the same RTT. To do this, the user must provide two parameters: the target RTT from C to S and the target data rate for the application. If users do not know these parameters, the default values, 90 Mb/s over a 20 ms path, are appropriate for most university users. The report presents various metrics of the local client and local path, and indicates if they are within the thresholds of TCP performance models. It also suggests corrective action, if needed.

The components of the *pathdiag* server are shown in Figure 2. The browser loads the diagnostic client, which communicates with the server via a simple request-response control protocol. A TCP connection is established from the traffic receiver in the diagnostic client to the traffic generator. The measurement engine uses the Web100 prototype of the extended statistics MIB [16] to manipulate and instrument the TCP connection at the generator. An analysis engine evaluates the measurements and extrapolates the results to predict the impact of the local path on the user's application.[2]

2.1   The Measurement Engine

The measurement engine collects Web100 data in a series of sample intervals. For each interval, it adjusts the window size of the diagnostic TCP connection in discrete steps, and then captures the entire set of Web100 variables at the end of each sample. It computes several metrics during each test, the most important of which are *DataRate*, *LossRate*, *RTT* and *Power* (*DataRate/RTT*). These are shown as functions of the window size for a typical link in Figure 3. These plots resemble those generated by "Windowed Ping" (*mping*) [14], a UDP-based tool that uses a similar measurement algorithm.

The measurement engine employs an adaptive scanner to select the window size for each sample interval. To minimize the total time required for the test,

---

[1] We use "target" when referring to components of the remote application and their parameters, such as end-to-end *target RTT*, desired *target data rate* and their product, the *target window size*.

[2] To support systems that cannot run a Java–enabled web browser, the "C" source for a portable command-line client is also published by the diagnostic server.

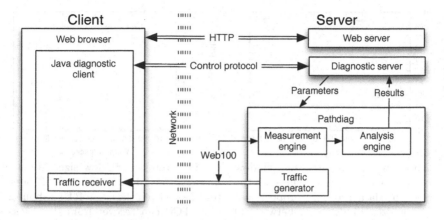

Fig. 2. Block diagram of the *pathdiag* client-server framework

data is collected in multiple phases that emphasize specific properties of the network. A coarse scan across the entire window range is used to approximately locate two important window sizes: the onset of queuing and the maximum window size. Ranges around these values are then rescanned at progressively higher resolutions. In Figure 3, the fine scans can be seen clearly around window sizes of 30 and 80 packets, respectively. The maximum window sizes for scans are determined when TCP congestion control or an end-host limitation prevents the window from rising for three consecutive sample intervals.

Several network path metrics are calculated directly from the raw data as it is collected. $MaxDataRate$ and $MinRTT$ yield a measurement of the test path's bandwidth-delay product. $MaxPowerWindow$ is the window size with the maximum $Power$, indicating the onset of queuing. The $MaxWindow$ is the maximum amount of unacknowledged data that the network held. The difference between the $MaxWindow$ and $MaxPowerWindow$ is an estimate of the queue buffer space at the bottleneck.

$BackgroundLossRate$ is calculated from the total packet losses from all sample intervals below the onset of queuing, as indicated by the $MaxPowerWindow$. It reflects bit errors and other losses that are not related to network congestion. If the adaptive scans do not provide sufficient loss data for the test described in the next section, additional loss data is collected at a fixed window size just below the onset of queuing. In general, the measurement engine collects enough data to observe the loss rate at the scale needed by AIMD congestion control to reach the target window size.

2.2 The Analysis Engine

The analysis engine uses the two user-supplied parameters, end-to-end RTT and desired application data rate to evaluate the results from the measurement engine and produce a diagnostic report, as shown in Figure 4.

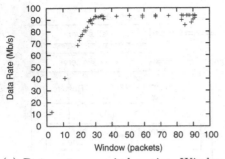

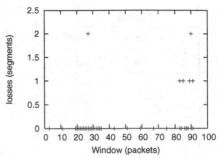

(a) Data rate vs. window size. Window sizes less than 30 were too small to fill the path, so the data rate was proportional to window size. Window sizes between 30 and 80 packets show data rates that were near the bottleneck rate, about 94 Mb/s.

(b) Loss rate vs. window size. Above 80 packets, the link started to exhibit persistent loss. Given the small RTT (about 2.5 ms), TCP can recover from these losses with only a slight reduction in throughput.

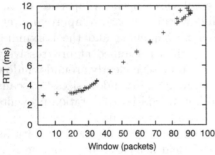

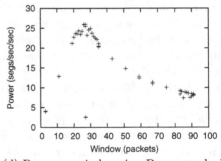

(c) RTT vs. window size. RTT was essentially constant at small window sizes. Above a window of 30, each additional packet in TCP's window was added to a standing queue at the bottleneck, and the RTT increased linearly with window size.

(d) Power vs. window size. Power reached a maximum at the point where the bottleneck crossed over from under-full (the link had idle time) to over-full (there was standing data in the queue), in some sense the ideal TCP operating point.

Fig. 3. Plots of scan results

The results in the generated report are grouped hierarchically. The base of the report shows test parameters and conditions. Test results are grouped into the following categories: local host (client) configuration, path measurements, and tester (server) consistency checks. Path measurements are further divided into data rate, loss rate, network buffering, duplex mismatch tests, and suggestions for alternate test parameters.

Test results are labeled and color-coded for easy reading. All failing tests (red) include a "corrective action" (starting with ">") indicating what needs to be fixed and how to fix it. In general, failing tests are guaranteed to be performance show-stoppers - the application will fail to meet the target data rate over the full end-to-end path as long as there are failing tests. The help for

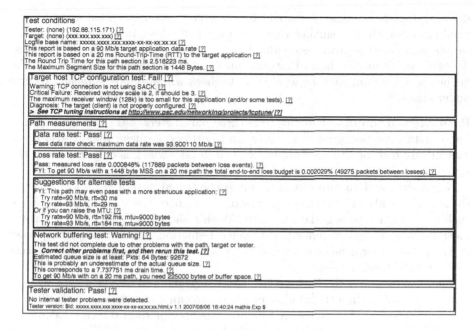

Fig. 4. Sample report from the same data as Figure 3

passing tests (green) indicates any caveats about limitations of the tests. Tests that are inconclusive for some reason yield orange warning messages. These include flaws that might not cause performance problems and tests that did not complete due to other failing tests. Messages in black are informational and are of most value to expert users. The analysis engine can detect 21 different failure conditions and 16 possible warnings.

Host Configuration. The host configuration tests confirm that TCP settings on the client are appropriate for target parameters. *Pathdiag* checks the options negotiated on the SYN and SYN-ACK. The Window Scale option [13] must have negotiated an appropriate value or it is flagged as a critical failure. It also checks if TCP Selective Acknowledgments (SACK) [18] or TCP Timestamps [13] are enabled.

A key test is whether the TCP receive buffer is larger than the target window. Since many modern operating systems adaptively size their receive buffers [9, 5, 23], it is necessary to check the announced receive window at extreme points of the measurements (peak data rate or window size). There are several corner cases that the analysis engine needs to consider. If the flow is limited by the receive window, and the maximum observed receive window for the entire run is less than the target window then the receiver never announced enough buffer space for the path, and the buffer is too small. If the receiver reduced its window at the extreme points, then receiver is not fast enough, which is a different problem. Also, since the section of the path under test is normally shorter than the target path, it might be correct for an adaptive receive window to have a maximum size that is smaller than the target window. In this case, *pathdiag* cannot make a strong conclusion about the receive buffer size. However, in most default configurations, hosts that announce sufficient receiver window for the queue space test and pass the window scale check will also have sufficient receive buffer space.

Path Measurements. *Pathdiag* tests three parameters of the local path: maximum data rate, background loss rate, and bottleneck queue size. It also has a special-case test for Ethernet duplex mismatch [21], which is only invoked if the signature is detected.

Normally, the data rate test fails only if the tested path is not short enough, the user is mistaken about the properties of the path, or there is a serious problem such as a media-type negotiation failure. With a short RTT, TCP can overcome most flaws with only a minor performance reduction. As a consequence, most flaws do not mask other flaws when the path is short enough, so a single test run can detect multiple flaws.

Pathdiag measures the background, non-congested loss rate at a window that is slightly smaller than the window necessary to cause congestion on the link. A failure is reported if the measured loss rate is greater than the rate calculated by a TCP performance model [19] applied to the specified RTT and data rate for the target application.

A warning is issued if the measured bottleneck queue buffer space is less than the bandwidth-delay product of the target path. It is only a warning because *pathdiag* cannot determine if the small buffer will cause significantly reduced performance for the target path. The test reflects the traditional full-BDP rule for sizing router buffers for TCP [22]. Recent results show that smaller buffers are adequate for aggregated flows [1,8], but for single flows there are some situations that might cause full window-sized bursts. In particular, TCP slow-start naturally requires $cwnd/2$ buffer space at the bottleneck to avoid prematurely transitioning to congestion avoidance [6]. Furthermore, if the bottleneck employs active queue management (AQM) [2] such as RED [7], *pathdiag* is likely to measure the threshold for dropping packets rather than the actual buffer space used to absorb bursts. We are planning future work in this area.

Tester Consistency Checks. Occasionally, either the traffic generator or *pathdiag* itself might be a bottleneck. For example, there may be unanticipated users on the server. *Pathdiag* checks for this and other exceptional events, and reports them as problems with the tester.

2.3   The Server Framework

The reports generated by *pathdiag* are ordinary web pages. They can be bookmarked and the URL forwarded to experts for additional analysis. The on-line documentation stresses this feature [15]. Even relatively naïve users can generate diagnostic reports that clearly identify a problematic subsystem, and then forward them to people with the resources and authority to take corrective action.

Web archival of *pathdiag* reports is also critical to our ongoing improvement of the tool. We periodically scan various deployed diagnostic servers and retrieve the reports from the analysis engine and the raw data from the measurement engine. We inspect selected reports to confirm they agree with our manual analysis of the raw data. If we discover flaws that are not reported clearly, we make improvements to the analysis engine. We test the improved analysis engine by reapplying it to our collection of measurement data, and inspect the re-generated reports that differ from the original reports. In this manner every user contributes to our pool of test data, and to refinement of the tool. Our archive currently holds more than 7000 diagnostic reports.

The sever does not expose any private information about the user except the name and IP address of the client machine. No user information or system version information is explicitly exposed, though some operating systems may be deduced by their performance properties.

3   Strengths and Weaknesses

Symptom scaling makes traditional tools currently used for network diagnostics, such as *ttcp* and *iperf*, completely insensitive to flaws on short paths. Network experts use these tools over long paths to test for the existence of flaws, but actually locating the flaws is often a difficult trial and error process. As described above, *Pathdiag*'s defining characteristic is its ability to compensate for symptom scaling. As such, it works best when run on short path sections, and is most useful for debugging problems close to end systems.

Pathdiag fundamentally relies on active measurement[3] and must send a significant amount of bulk data to measure the loss rate at the scale of the target application. In this way it is different than many bandwidth-estimation tools that obtain results by measuring dispersion of short bursts of traffic without sending sufficient traffic to measure the loss rate at a scale relevant to AIMD congestion control.

The measurement algorithms used by *pathdiag* assume that other traffic across the tested path is relatively unvarying. Though it will not result in a false pass, highly variable levels of cross traffic may yield inconsistent results, especially in measurements of bottleneck buffer size and measured throughput. This can be largely mitigated by testing a shorter section of the path.

One fairly basic limitation is that the underlying diagnostic TCP stream is unidirectional, and TCP is intrinsically difficult to instrument from the receiving end. There are a number of potential solutions to this, which we hope to address in future work. Users *can* test the reverse path by running *pathdiag* at the other end of the test path. This can be done most easily by using the Internet2 Network Performance Toolkit [3] live boot CD to run a temporary server on almost any PC.

---

[3] For specialized uses, *pathdiag* can be run from the command line as a standalone tool without the web server server framework. One special use is to manually attach it to a bulk TCP stream belonging to another application.

Pathdiag cannot diagnose application problems, since the target application does not participate in the testing process. It is often very difficult to write applications that can attain high data rates even on ideal long networks. Some application problems are addressed in related work [10, 20].

4   Closing

Pathdiag is designed to improve TCP performance for the Research and Education masses—those with a need for high performance but without the time or expertise to individually diagnose network problems. It is particularly well suited for testing at the edges of the network, which is usually where the majority of performance-reducing flaws occur. Since it compensates for symptom scaling, *pathdiag* is able to isolate these near-edge flaws that are very difficult to diagnose using conventional local diagnostics.

In its most common form, deployment of *pathdiag* is fairly straightforward. A single well-connected test server is in a position to provide coverage for an entire campus or metropolitan network. It is our intent that *pathdiag* test servers will ultimately yield significant benefits to both the users and administrators of high-performance networks.

References

1. Appenzeller, G., Keslassy, I., McKeown, N.: Sizing router buffers. In: Proc. of ACM SIGCOMM 2004, October 2004, pp. 281–292 (2004)
2. Braden, B., et al.: Recommendations on queue management and congestion avoidance in the internet. In: RFC 2309 (April 1998)
3. Carlson, R.: Network performance toolkit,
   http://e2epi.internet2.edu/network-performance-toolkit.html
4. Carpenter, B., Brim, S.: Middleboxes: Taxonomy and issues. In: RFC 3234 (February 2002)
5. Fisk, M., Feng, W.: Dynamic right-sizing is TCP. In: 2nd Annual Los Alamos Computer Science Institute Symposium (LACSI 2001) (October 2001)
6. Floyd, S.: Limited slow-start for TCP with large congestion windows. In: RFC 3742 (March 2004)
7. Floyd, S., Jacobson, V.: Random early detection gateways for congestion avoidance. IEEE ACM Transactions on Networking 1(4), 397–413 (1993)
8. Ganjali, Y., McKeown, N.: Update on buffer sizing in internet routers. ACM CCR 36(4), 67–70 (2006)
9. Heffner, J.: High bandwidth TCP queuing,
   http://www.psc.edu/~jheffner/papers/senior_thesis.pdf
10. Heffner, J., Mathis, M.: Applications and the speed of light: How well do applications perform on long perfect networks (2007), Web paper:
    http://www.psc.edu/networking/projects/applight/
11. Internet2 Land Speed Record, http://www.internet2.edu/lsr/
12. Internet2 NetFlow Weekly Reports, http://netflow.internet2.edu/weekly/
13. Jacobson, V., Braden, B., Borman, D.: TCP extensions for high performance. In: RFC 1323 (May 1992)

14. Mathis, M.: Windowed ping: an IP layer performance diagnostic. Computer Networks and ISDN Systems 27(3), 449–459 (1994)
15. Mathis, M., et al.: NPAD diagnostics servers: Automatic diagnostic server for troubleshooting end-systems and last-mile network problems (2007), Web paper: http://www.psc.edu/networking/projects/pathdiag/
16. Mathis, M., Heffner, J., Raghunarayan, R.: TCP extended statistics MIB. In: RFC 4898 (May 2007)
17. Mathis, M., Heffner, J., Reddy, R.: Web100: Extended TCP instrumentation for research, education and diagnosis. Computer Communications Review 33(3), 69–79 (2003)
18. Mathis, M., Mahdavi, J., Floyd, S., Romanow, A.: TCP selective acknowledgement options. In: RFC 2018 (October 1996)
19. Mathis, M., Semke, J., Mahdavi, J.: The macroscopic behavior of the TCP congestion avoidance algorithm. Computer Communications Review 27(3), 67–82 (1997)
20. Rapier, C., Stevens, M.: High performance SSH/SCP - HPN-SSH (2007), http://www.psc.edu/networking/projects/hpn-ssh/
21. Shalunov, S., Carlson, R.: Detecting duplex mismatch on ethernet. In: Dovrolis, C. (ed.) PAM 2005. LNCS, vol. 3431, pp. 135–148. Springer, Heidelberg (2005)
22. Villamizar, C., Song, C.: High performance TCP in ANSNET. Computer Communications Review 24(5), 45–60 (1994)
23. New networking features in Windows Server 2008 and Windows Vista (2008), http://technet.microsoft.com/en-us/library/bb726965.aspx

SCUBA: Focus and Context
for Real-Time Mesh Network Health Diagnosis

Amit P. Jardosh, Panuakdet Suwannatat, Tobias Höllerer, Elizabeth M. Belding,
and Kevin C. Almeroth

Department of Computer Science, UC Santa Barbara

Abstract. Large-scale wireless metro-mesh networks consisting of hundreds of
routers and thousands of clients suffer from a plethora of performance problems.
The sheer scale of such networks, the abundance of performance metrics, and the
absence of effective tools can quickly overwhelm a network operators' ability to
diagnose these problems. As a solution, we present *SCUBA*, an interactive *focus
and context* visualization framework for metro-mesh health diagnosis. SCUBA
places performance metrics into multiple tiers or *contexts*, and displays only the
topmost context by default to reduce screen clutter and to provide a broad contextual overview of network performance. A network operator can interactively
focus on problem regions and zoom to progressively reveal more detailed contexts only in the focal region. We describe SCUBA's contexts and its *planar* and
hyperbolic views of a nearly 500 node mesh to demonstrate how it eases and ex-
pedites health diagnosis. Further, we implement SCUBA on a 15-node testbed,
demonstrate its ability to diagnose a problem within a sample scenario, and discuss its deployment challenges in a larger mesh. Our work leads to several future
research directions on focus and context visualization and efficient metrics collection for fast and efficient mesh network health diagnosis[1].

Keywords: wireless mesh networks, network visualization, network health.

1 Introduction

Metro-scale wireless mesh networks (WMNs)[2], consisting of hundreds of routers, are
being deployed worldwide in city downtowns, malls, and residential areas[3]. While several millions of dollars have been spent to deploy WMNs, these networks suffer from
a plethora of problems that severely impact their performance. Some of the most common problems are weak client connectivity due to signal attenuation, interference from
external devices, and misbehaving or misconfigured client nodes [1]. These problems
have largely been responsible for WMN vendors not achieving sustainable client market
penetration, thereby leading to dwindling business prospects for this technology.

We believe that the effective diagnosis and troubleshooting of performance problems is key to the success of metro-scale WMNs. Although many novel metrics and
techniques to diagnose and troubleshoot problems in WMNs have been proposed by

---

[1] A video demo of SCUBA is at http://moment.cs.ucsb.edu/conan/scuba/

[2] http://www.muniwifi.org/

[3] www.tropos.com,www.firetide.com,www.strixsystems.com,www.meraki.com

M. Claypool and S. Uhlig (Eds.): PAM 2008, LNCS 4979, pp. 162–171, 2008.
© Springer-Verlag Berlin Heidelberg 2008

the research community [9,6], sifting through a sea of such metrics collected from each device in a metro-scale WMN can be overwhelming for network operators.

As a solution, diagnostic tools utilize visualization techniques such as time-series plots and planar graphs[4]. However, the diagnosis of problems by viewing a myriad of such graphs and plots in large-scale WMNs can be very tedious and time-consuming. We believe that operators of large-scale WMNs need clever *structured* visualization techniques to quickly navigate through metrics and diagnose problems. Numerous publications have shown that good visualizations decrease the time and effort to evaluate large volumes of information in the Internet [11,10,8][5]. To our knowledge, diagnostic visualizations of large-scale WMNs have received little to no research attention yet. In this paper we argue that these networks can certainly benefit from visualization tools, especially due to their increasing sizes and complexities.

To this end, we propose a *focus and context* visualization framework named *SCUBA*[6]. SCUBA places performance metrics into several tiers or *contexts*. The topmost context provides a WMN operator with a broad contextual overview of WMN performance. By viewing only this broad context, WMN operators can quickly identify and locate problems within the WMN. Once a problem location is determined, an operator can choose to narrow his/her focus on the problem region and zoom to reveal detailed metric contexts within that region. In other words, the operator exposes a larger set of metrics within a small focal region to diagnose the cause of a performance problem.

In this paper we propose a scheme for organizing metrics into three contexts (route, link, and client) with increasing detail. The placement of metrics is based on our experience of diagnosing WMN problems [5]. However, the main objective of SCUBA is to *facilitate* focus and context visualization for any scheme. Different schemes derived from WMN operators' common diagnostic approaches will be explored in the future to define better contexts as well as better placement of metrics within contexts.

To explain contexts, metrics, and views of SCUBA, we utilize the Google Mountain View WMN map of about 500 routers and gateways[7]. To understand SCUBA's ease of use in diagnosing a sample performance problem and it deployment challenges, we implement it on the 15-node UCSB MeshNet [5].

2   *SCUBA*: Focus and Context Visualizations

The main objective of SCUBA is to facilitate fast and easy diagnosis of WMN performance problems by cleverly organizing the performance metrics for focus and context visualizations. In this section we discuss the metrics collection architecture, the organization of metrics into contexts, the different views SCUBA offers to the operator, and the variety of visualization features implemented in SCUBA.

---

[4] NetCrunch: http://www.adremsoft.com/netcrunch/index.php

[5] CAIDA tools: http://www.caida.org/tools/visualization; NetDisco: http://www.netdisco.org

[6] The name SCUBA comes from the sport of scuba diving, where a diver swims close to the water surface and dives deeper to get a closer look at what is beneath the surface.

[7] http://wifi.google.com/city/mv/apmap.html

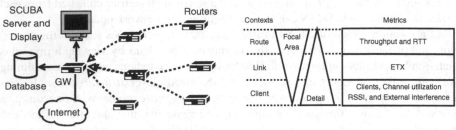

(a) SCUBA's metric collection architecture.   (b) Metric contexts used in this work.

Fig. 1. SCUBA's metrics collection architecture and metric contexts

2.1 Metrics Collection Architecture

The performance metrics visualized by SCUBA are collected and computed from the routers and gateways in a WMN. As shown in Figure 1(a), each router sends a set of metrics to the SCUBA server via the gateway. The SCUBA server stores these metrics in two locations: a database so that temporal trends of metrics can be observed, and a data structure within main memory for fast access by SCUBA's visualization engine. The SCUBA visualization engine is a standalone Java application written using the Swing GUI toolkit. We discuss the computation and collection of metrics specifically within the UCSB MeshNet in Section 3.

2.2 SCUBA's Contexts

One of the main obstacles to diagnosing problems in WMNs today is their sheer scale and the abundance of performance metrics that can be overwhelming to the WMN operator and unrealistic to analyze within a short period of time. To better organize the collected information, we propose that WMN performance metrics be placed into several *contexts*, where each context consists of one or more metrics.

The topmost context provides the WMN operator with a holistic view, a broad contextual overview of WMN health. In other words, a WMN operator can quickly identify both the occurrence and the location of a problem in the WMN from such a broad context. An operator can then use SCUBA to *focus* on specific problem areas. Once the operator shifts focus, he/she can interactively zoom to view more detailed contexts. In other words, the operator can choose to reveal more metrics and therefore increase information detail *isolated within the focal area*.

For the scope of this paper, we place WMN metrics within three contexts; the *route*, *link*, and *client* contexts. These three contexts and their metrics are summarized in Figure 1(b). The figure shows that as information detail increases, the focal region is narrowed in the lower SCUBA contexts. We next describe the organization of metrics in the three contexts and explain how these metrics help isolate causes of a sample WMN performance problem.

Route Context: The route context is the topmost context and only displays multi-hop routes between each router and its corresponding Internet gateway. The context consists of two metrics: (a) throughput of TCP flows over the multi-hop routes formed from each router; and (b) the round-trip time (RTT) of UDP packets on the same routes. The two metrics are computed by each router and determine the quality of the route between the router and its gateway. We include these metrics in the topmost context of SCUBA because any significant drop in their values indicates a serious performance problem. WMN operators can use this problem indication and then zoom into the problem region to understand the real cause of a problem. For instance, problems such as sudden route flaps, unexpected drop in throughput, or an increase in RTT values can cause a performance deterioration of TCP or UDP application flows that utilize those routes. Operators can further investigate the cause of such problems by increasing the context in the problem areas.

Link Context: The link context reveals one additional metric, the expected transmissions count (ETX) [3] on a link between the nodes. In the link context, SCUBA displays the point-to-point MAC-layer links between nodes in addition to the routes from the route context. We use ETX as a metric in this context because it provides a good estimate of the health of links between nodes. The quality of links is likely to impact the routes that utilize them. As a result, if sudden route flaps or a significant drop in throughput are observed at the route context, the most likely cause is poor quality links utilized by the routes. Poor link quality is identified by an increase in the ETX value at the link context, and typically occurs due to three reasons: (a) heavy volume of traffic flowing over the link and/or neighboring links within its interference region; (b) external interference from a co-located radio wave source that does not belong to the WMN; and (c) heavy signal attenuation caused by some obstacle. Isolation of the causes of poor links is achieved by zooming to the next lower context.

Client Context: The client context provides further insight into the cause of poor quality links. SCUBA includes four metrics within this context: (a) the number of clients associated with each router; (b) the percentage channel utilization per client [4]; (c) the received signal strength indicator (RSSI) of MAC-layer frames received from clients; and (d) the volume of external interference. These metrics are included within this context because they each describe client connections and traffic within a WMN. In the client context, SCUBA displays the clients associated with the routers, along with the links and the routes from the link context. A WMN operator will likely zoom to the client context only when the cause of problems cannot be easily determined at the link context. For instance, the cause of poor link quality can be isolated to either a large number of clients with high channel utilization values or external interference[8]. Both these causes can be determined from metrics in the client context. If neither have adversely impacted the quality of links, the WMN operator can determine that heavy signal attenuation by an obstacle is the likely cause of poor quality links, by the process of elimination.

2.3 Diagnostic Approaches and the Design of SCUBA

The three contexts and the placement of metrics within the contexts we present for the current version of SCUBA have been designed based on our own experience of

---

[8] We compute external interference as the percentage of channel utilized by transmitters that are not associated with a router.

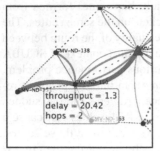 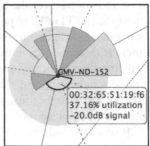 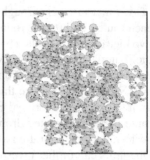

(a) Route throughput, RTT and link ETX.   (b) Client channel utilization and signal strengths.   (c) Google WMN, without SCUBA's focus and context.

Fig. 2. SCUBA's visualization features and an example WMN without focus and context

building and deploying a WMN [5], and diagnosing problems using a logical *top-to-bottom* approach. In the future, we plan to evaluate additional contexts such as *traffic* and *application*, for increased diagnosis flexibility in specific application settings. While our scheme is sufficiently general for diagnosing a wide variety of problems [1], it does not represent an all-inclusive set of metrics or the only scheme of context organization. WMN operators may follow alternative diagnostic approaches in different deployment scenarios, and the metrics they may find useful in each scenario can also vary. SCUBA, as a visualization framework, can be modified to utilize different schemes based on the diagnostic approaches preferred by operators. The effectiveness of SCUBA should then be evaluated qualitatively and quantitatively in specific scenarios, using metrics such as its ease of use, how quickly it can help diagnose a problem, and how many problems of interest it helps diagnose. Exploring other diagnostic approaches and evaluating their effectiveness, while outside of the scope of this paper, is part of our usability-oriented ongoing work.

2.4 SCUBA Visualization Features

In this section we discuss the visualization features we use to communicate WMN health using the seven metrics discussed in the previous section. We use different color and size schemes for these features with a single consistent visualization policy across all contexts and metrics, which is to *highlight problems* in the WMN, resulting in fast and easy diagnosis of WMN problems. SCUBA's visualizations are interactive, allowing for continuous pan and zoom and tool-tip-style data readouts on mouse-over and selection. The visualization features, as illustrated in Figure 2, are as follows:

WMN Nodes: We assume that a typical WMN backbone consists of two types of nodes: routers and Internet gateways. SCUBA displays routers as blue circles and gateways as more salient red triangles, as shown in Figure 2 and 3.

Routes: WMN routers relay client packets destined for the Internet via other routers towards a gateway. The gateways relay packets destined for WMN clients towards the router with which they are associated. SCUBA visualizes the routes between routers and gateways as curved solid lines, as illustrated in Figure 2(a). In order to implement our policy of highlighting problems, the thickness of the lines is directly proportional to the RTT value; the higher the RTT, the thicker the line. On the other hand, the saturation and brightness levels of the line is inversely proportional to the throughput on an exponential scale using the HSB color scheme; low throughput routes appear bright red, while higher throughput routes are de-emphasized with a grey color. We also experimented with a more conventional mapping of throughput to line thickness, which might be preferable for non-troubleshooting monitoring applications, but the presented scheme is advantageous when high salience of trouble spots is important.

Links: MAC-layer links between nodes are visualized at the link context. The links are visualized as dashed lines, as illustrated in Figure 2(a). To maintain our policy of highlighting problems, the length of white spaces between dashes are directly proportional to the ETX value; higher the ETX value, longer are the white spaces, the more *broken* the links appear. In order to make up for the reduction in saliency by the increasing gap sizes, the thickness of the dashed lines are increased proportional to the ETX value. This visualization feature ensures that the operators' attention is drawn to poor quality broken links, and less towards good quality links represented as thin solid lines.

Clients: The client context of SCUBA shows clients and four related metrics. These metrics are illustrated in Figure 2(b). In this figure, the clients are placed around the router with which they are associated, and are visualized as sectors of a circle. The subtended angle of the client's sector is a value between 0° and 360°, proportional to the client's percentage channel utilization share. As a result, a router with client sectors that form a complete circle has its entire (100%) channel utilized by client frame transmissions. The radius of each client sector is inversely proportional to the RSSI value of the client's frames received by the router. As a result, the lower the client's RSSI, the farther the client is placed from the router, and the larger the radius. Based on these two visualization features of client sectors, a client with a large sector angle and large radius is quickly seen as a potential problem because of high channel utilization and low RSSI. The client with the largest sector area is highlighted in a bright green color, making it easy for a WMN operator to locate all problem clients at the client context. The fourth metric, external interference, is visualized as a grey cloud around routers, as shown in Figure 2(c). The radius of the cloud is directly proportional to the volume of external interference. Moreover, when the interference cloud of two or more routers overlap, SCUBA darkens the color in the region of overlap, indicating more interference.

2.5 SCUBA Views

In this section, we discuss two views of SCUBA, *planar* and *hyperbolic*. These views further ease the diagnosis of problems in large-scale WMNs by facilitating *focus and context* interaction [2] of the WMN operator with SCUBA's contexts. In other words, using either of the two views operators can choose to focus on a specific location in the view while retaining some kind of overview of the whole network, and they can zoom to a context of their choice for further investigation of problems.

To understand the benefit of these views, we use the Google WMN in Mountain View, California, which consists of 425 routers and 66 gateways. Since we do not have access to the actual metrics from this WMN, we use the geo-locations of the routers

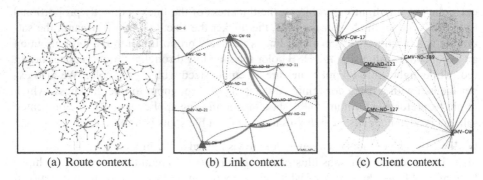

(a) Route context.   (b) Link context.   (c) Client context.

Fig. 3. Google WiFi mesh network using SCUBA's planar view

to create a sample large-scale visualization environment and synthetically generate values for metrics using simple assumptions. Links between routers and their corresponding ETX values are generated in loose correlation with the physical distances between routers. Routes are computed using a shortest path algorithm between the routers and their closest gateways. The throughput and RTT values are computed based on cumulative ETX values of links utilized by the routes to the gateway. Zero or more clients are matched with routers, such that their total count loosely approximates those published in a recent news article presenting statistics on the Google WMN [1]. The channel utilization, signal strength, and external interference metrics are randomly chosen from a uniform distribution.

To clearly demonstrate the advantages of the focus and context visualizations of SCUBA, in Figure 2(c) we show a screenshot of all seven metrics from each of the three contexts displayed for the Google WMN. Because of the size of network, the screenshot appears cluttered, thereby limiting the ability of an operator to extract any coherent information from the view for problem diagnosis. We now discuss the two interactive SCUBA views, how they reduce screen clutter, their advantages over each other, as well as their trade-offs.

Planar View: SCUBA's planar view is shown in Figure 3. The WMN and its several contexts are rendered on a flat two-dimensional plane. Figure 3(a) shows the planar view with only the route context displayed for the entire Google WMN. Figure 3(b) shows the link context of a small subset of the network, when the WMN operator zooms to investigate any performance problems identified at the route context. These figures also show an *inset overview* in the top-right corner that indicates the focal region in the overall view. Further zooming reveals the client context, illustrated in Figure 3(c). The focus region in the overview inset is seen to shrink in size, because the operator is now zoomed to a smaller focus region.

The advantage of SCUBA's planar view is that it maintains the geographical location and orientation of all the routers and gateways, even while an operator changes focus and context. However, the trade-off of planar views is that while operators are zoomed in on lower metric contexts, they can only see the small inset overview of the whole

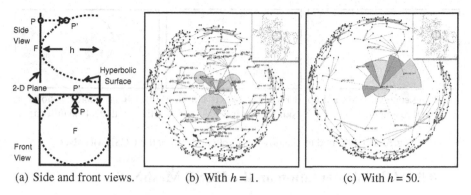

(a) Side and front views.   (b) With $h = 1$.   (c) With $h = 50$.

Fig. 4. Google WiFi mesh network using SCUBA's hyperbolic view

network, which may not be sufficient to alert them to possible new unusual activity. We overcome this problem by using *hyperbolic* views [7].

Hyperbolic View: SCUBA uses the hyperbolic view to render routers, gateways, clients, and their corresponding metrics on a hyperbolic surface [7]. The basic idea of a hyperbolic view is to plot the focal point F of a two-dimensional plane at the center of the screen, and plot the remaining points on a hyperbolic surface, centered at the focal point. Figure 4(a) illustrates the side and front of a hyperbolic surface transformed from a simple two-dimensional planar surface. The figure shows that the non-focal location point P on a two-dimensional plane is distorted to P' when transformed to the hyperbolic surface. The distortion depends on the height h of the hyperbola.

The hyperbolic view has an advantage over the planar view in that it *automatically* renders different contexts of SCUBA based on the position of the node with respect to the focal point. As shown in Figure 4(b), the node of interest forms the focal point of the hyperbolic surface and the remaining nodes are rendered on the hyperbolic surface, using the same *orientation* to the focal node as in the planar view. Also, as illustrated in Figure 4(b), SCUBA plots all the metric contexts for the focal node and progressively reduces the contexts for nodes further away from the focal node. As a result, only the route context is displayed for the nodes at the edge of the hyperbolic surface. Figure 4(c) shows that a parameter controls the depth of the hyperbolic surface, which determines how quickly context displays are reduced between the focus node and the surface edge.

A main advantage of the hyperbolic view is that it shows a complete view of the WMN at all times, and automatically changes contexts as the operator interactively changes focus by mouse-dragging. As a result of this automation, the user is required to only choose his/her focus point, and SCUBA smoothly transitions to display the new focal region and the corresponding contexts. However, the trade-off of hyperbolic views is that it distorts the geographic locations of the nodes from that of a planar view. As a result of these trade-offs between the two views, SCUBA includes an inset overview similar to the one used in the planar view and allows a WMN operator to quickly toggle between the planar and hyperbolic views.

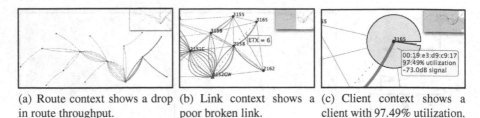

(a) Route context shows a drop in route throughput.

(b) Link context shows a poor broken link.

(c) Client context shows a client with 97.49% utilization.

Fig. 5. SCUBA used to diagnose a sample problem in the UCSB MeshNet

3   SCUBA Implementation on the UCSB MeshNet

Our main goal for SCUBA is to make it easily usable and effective in diagnosing practical WMN problems. To test this capability, we have used SCUBA to study the performance of the UCSB MeshNet, an indoor WMN consisting of 14 multi-radio 802.11 a/g routers and one gateway. The routers collect, compute, and send metrics to a central SCUBA server via the gateway over the routes. In this section, we show how SCUBA is used to diagnose a sample performance problem within the MeshNet.

In the route context shown in Figure 5(a), the operator observes the unusually low throughput and high RTT values of a route indicated by the thick red lines. The operator zooms in the region of the route to access the link context. The link context in Figure 5(b) clearly shows that the links utilized by the problem route have a high ETX value, indicated by the sparsely dashed straight lines. To determine the cause, the operator zooms to the client context close to the edge router using the problem route, as shown in Figure 5(c). In this figure, the large sector of the circle representing a client with 97.49% channel utilization looks clearly anomalous. The operator is thus assured that a single misconfigured and/or misbehaving client is overloading the channel with excess traffic and adversely impacting the performance of an entire route.

The current version of SCUBA allows operators to diagnose several other performance problems, such as a flashcrowd of users overloading the network or suboptimal route topologies caused due to poor links or interference. The set of diagnosable problems will increase with the number and type of metrics collected. Moreover, the addition of a *time* dimension will allow SCUBA to diagnose many more temporal problems, such as rapid route flaps and client mobility.

4   Conclusions

In this paper, we propose a focus and context visualization framework called SCUBA for fast and efficient WMN health diagnosis. SCUBA places WMN performance metrics into contexts and presents them in two views, planar and hyperbolic. We believe that visualization frameworks such as SCUBA will form the most structured and efficient means of WMN health diagnosis.

SCUBA opens several new directions of research. The most prominent one is to qualitatively and quantitatively determine the *best* set of metrics and contexts that facilitate comprehensive diagnosis. To do so, we believe that qualitative usability studies of SCUBA's visualization methods and the study of various diagnostic approaches preferred by WMN operators will be very helpful. Another research direction is to reduce the metrics' computation and collection overhead to achieve *real-time* visualization capabilities. A possible future extension is to make SCUBA use a set of diagnostic rules to *automatically* identify problem regions and adjust focus and context accordingly. Such automation will immediately direct an operator's attention to the problem and will likely reduce diagnosis time.

SCUBA is the first step towards interactive visualizations for fast and efficient WMN health diagnosis. We believe that as WMNs are rapidly deployed worldwide and as they increase in complexity, the need for such visualizations will grow. Faster and efficient health diagnosis will help operators maintain their WMN's performance and therefore achieve the desirable economic success of the metro-scale mesh technology.

Acknowledgments

This work was funded in part by NSF Wireless Networks (WN) award CNS-0722075.

References

1. Tropos Report on Google WiFi Network,
   www.muniwireless.com/article/articleview/5403
2. Card, S.K., Mackinlay, J.D., Shneiderman, B.: Readings in Information Visualization: Using Vision to Think. Morgan Kaufmann Publishers Inc, San Francisco, CA
3. De Couto, D., Aguayo, D., Bicket, J., Morris, R.: A High-throughput Path Metric for Multi-hop Wireless Routing. Wireless Networks 11(4), 419–434 (2005)
4. Jardosh, A.P., Ramchandran, K.N., Almeroth, K.C., Belding, E.M.: Understanding Congestion in IEEE 802.11b Wireless Networks. In: Proceedings of USENIX IMC, Berkeley, CA (October 2005)
5. Lundgren, H., Ramachandran, K.N., Belding-Royer, E.M., Almeroth, K.C., Benny, M., Hewatt, A., Touma, A., Jardosh, A.P.: Experiences from the Design, Deployment, and Usage of the UCSB MeshNet Testbed. IEEE Wireless Communications Magazine 13, 18–29 (2006)
6. Marti, S., Giuli, T., Lai, K., Baker, M.: Mitigating Routing Misbehavior in Mobile Ad hoc Networks. In: Proceedings of MOBICOM, Boston, MA, pp. 255–265 (2000)
7. Munzner, T.: Interactive Visualization of Large Graphs and Networks. PhD thesis, Stanford University (June 2000)
8. Paxson, V.: Strategies for Sound Internet Measurement. In: Proceedings of IMC, October 2004, pp. 263–271. Taormina, Sicily (2004)
9. Qiu, L., Bahl, P., Rao, A., Zhou, L.: Troubleshooting Wireless Mesh Networks. ACM SIG-COMM Computer Communication Review 36(5), 17–28 (2006)
10. Sommers, J., Barford, P., Willinger, W.: SPLAT: A Visualization Tool for Mining Internet Measurements. In: Proceedings of PAM, Adelaide, Australia (March 2006)
11. Tukey, J.: Exploratory Data Analysis. Addison-Wesley, Menlo Park, CA (1977)

IMR-Pathload: Robust Available Bandwidth Estimation Under End-Host Interrupt Delay

Seong-Ryong Kang and Dmitri Loguinov

Texas A&M University, College Station, TX 77843, USA
{skang,dmitri}@cs.tamu.edu

Abstract. Many paths in PlanetLab cannot be measured by Pathload. One of the main reasons for this is timing irregularities caused by interrupt moderation of network hardware, which delays generation of interrupts for a certain period of time to reduce per-packet CPU overhead. Motivated by this problem, we study Pathload in detail under various end-host interrupt delays and find that its trend detection mechanism becomes susceptible to non-negligible interrupt delays, making it unable to measure network paths under such conditions. To overcome this, we propose a new method called IMR-Pathload (*Interrupt Moderation Resilient Pathload*), which incorporates robust trend detection algorithms based on signal de-noising techniques and reliably estimates available bandwidth of network paths under a wide range of interrupt delays. Through experiments in Emulab and Internet, we find that IMR-Pathload substantially improves Pathload's measurement reliability and produces accurate bandwidth estimates under a variety of real-life conditions.

Keywords: Bandwidth estimation, network measurement, interrupt moderation, and interrupt delays.

1 Introduction

Bandwidth of Internet paths is an important metric for applications. Extensive research has been conducted over the years and the vast majority of work in this area focuses on end-to-end measurement. Although several techniques [4], [13], [11], [12], [14] attempt to measure capacity of the narrow link (i.e., the slowest link in a path) or both capacity and available bandwidth of the tight link (i.e., link with the smallest available bandwidth over a path), many measurement techniques and public tools (such as [6], [9], [16]) have been developed to estimate available bandwidth of the tight link. These methods mainly focus on fast estimation with high accuracy under a various traffic conditions. However, since the ultimate goal of bandwidth estimators is to measure diverse Internet paths, before being a full-blown measurement tool, it is highly desirable that tools are resilient to timing irregularities caused by various OS scheduling delay jitter or hardware interrupt moderation in real networks.

Note that to accurately measure bandwidth, all existing methods heavily rely on high-precision delay measurement of probe packets at end-hosts. However,

M. Claypool and S. Uhlig (Eds.): PAM 2008, LNCS 4979, pp. 172–181, 2008.

irregular timing due to interrupt moderation at network interface cards (NICs) has been identified as the major problem of existing bandwidth estimation tools in practice [15]. To reduce the effect of interrupt moderation, recent tools such as Pathchirp [16] and Pathload described in [15] incorporate mechanisms that aim to "weed out" packets affected by interrupt delays. However, Pathchirp requires manual modification to force it to send (often substantially) more probing packets to obtain an accurate estimate, prolonging measurement undesirably. On the other hand, Pathload attempts to filter out affected packets without increasing the number of probing packets, which unfortunately has a limited effect when interrupt delays become non-trivial. This makes Pathload's estimation much more susceptible to error, which happens fairly often in practice.

To address the above filtering problem without increasing measurement duration, we investigate Pathload's internal algorithm and find that its estimation instability with non-negligible interrupt delays stems from its delay-trend detection mechanism that is not robust under bursty packet arrival introduced by network hardware. To overcome this, we introduce two trend-detection algorithms based on signal de-noising techniques such as wavelet decomposition and window-based averaging and call the new method IMR-Pathload (*Interrupt Moderation Resilient Pathload*). Through experiments in Emulab [5] under various network settings, we find that IMR-Pathload significantly improves Pathload's performance in a wide range $(0 - 500 \ \mu s)$ of interrupt delays δ. Especially, under non-trivial interrupt delays (e.g., $\delta > 125 \ \mu s$), while Pathload fails to produce estimates for any of the paths studied in this paper, IMR-Pathload measures their available bandwidth with over 88% accuracy. Internet experiments also confirm that IMR-Pathload reliably produces bandwidth estimates even for the paths that are not measurable by Pathload.

2   Related Work

A number of techniques have been proposed to measure available bandwidth of network paths [6], [9], [16], which sends N back-to-back packets and discover a relationship between sending rates at the sender and the corresponding receiving rates at the receiver to produce bandwidth estimates of the paths. Among them, we discuss two promising tools that use mechanisms to mitigate the effect of interrupt moderation.

Pathchirp [16] uses packet-trains (called chirps) with exponentially decreasing inter-packet spacings in each chirp and infers available bandwidth using the queuing delay signature of arriving chirps. The basic idea behind this method is that when a transmission rate r_k of a packet k in a chirp reaches available bandwidth of a path under consideration, then subsequent packets $j > k$ in the chirp will exhibit increasing queueing delay. Hence, available bandwidth of the path is the rate r_k of the packet k whose queueing delay starts increasing. To overcome the packet-timing problem introduced by end-host interrupt moderation, Pathchirp increases the number of probing packets in each chirp by a manually selected amount and uses only those packets that (ideally) have not been affected by interrupt delays.

Different from Pathchirp, Pathload [9] sends a fleet of packet-trains with a fixed rate and adjusts the sending rate for the next fleet based on delay-trend information provided by the receiver. Pathload searches for an available bandwidth range by increasing or decreasing the sending rate of probe-trains in a binary search fashion according to trend information. Although Pathload can reduce the effect of interrupt delays without increasing the number of packets in each probe-train, its algorithm is effective only under small interrupt delays.

3   Issues of Interrupt Delay in Bandwidth Measurement

As use of interrupt moderation has become a common practice in modern network settings, host machines in real networks may employ interrupt delays that vary widely in order to reduce CPU utilization and to increase network throughput. It is reported in [7] that the range of interrupt delays recommended for Intel Gigabit NIC (GbE) is $83 - 250$ μs for Microsoft Windows-based systems and $125 - 1000$ μs for Linux-based systems. Jin et $al.$ [10] also report that a variety of systems equipped with Gigabit NICs require to delay generation of interrupts over 470 μs to achieve good throughput in receiving high-speed TCP streams and to substantially reduce CPU utilization. The question we have now is how this wide range of interrupt delays affects Pathload's bandwidth estimation. We discuss this issue next.

3.1   Impact of Interrupt Delay

To investigate the potential impact of interrupt moderation on Pathload, we conduct experiments in Emulab [5] for different interrupt delays at the receiver[1]. We start by describing the experimental setup.

Experimental Setup. For this investigation, we use a topology shown in Fig. 1, in which source PS sends probe data to the destination PR through five routers $R_1 - R_5$. Nodes S_i ($i = 1, 2, 3, 4$) send cross-traffic packets to destination nodes D_i at an average rate λ_i. The speed of all access links is 100 Mb/s (delay 10 ms) and the remaining links L_i ($i = 1, 2, 3, 4$) between routers R_i and R_{i+1} have capacities C_i and propagation delay 40 ms.

To examine Pathload's estimation reliability, we use six different network settings shown in Table 1, which lists the capacity C_i and available bandwidth A_i of each link L_i for different experimental scenarios. The shaded values in each row represent the tight-link capacity and available bandwidth of the path for each case. The values in square brackets represent the capacity of the narrow link (i.e., bottleneck bandwidth) for each case.

In all experiments, we use TCP cross-traffic generated by Iperf traffic generators [8] to load network paths. For this purpose, we run 100 threads in each cross-traffic source S_i to generate TCP flows that are injected into routers R_1, R_2, and R_3 and keep the utilization of each router R_i according to the values

---

[1] In Emulab, users can change configuration of network cards.

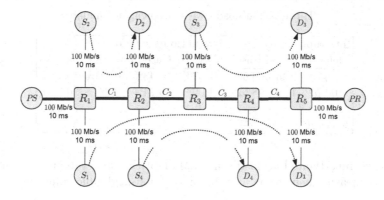

Fig. 1. Evaluation topology in Emulab

Table 1. Evaluation Setup

Experimentation scenarios	Different link bandwidths (Mb/s)			
	C_1 A_1	C_2 A_2	C_3 A_3	C_4 A_4
Case-I	75 31.84	90 51.69	90 42.05	[60] 40.77
Case-II	75 41.32	90 70.76	90 46.77	[60] 26.39
Case-III	[60] 35.88	90 70.76	[90] 23.39	75 18.10
Case-IV	[60] 21.60	90 65.99	90 42.07	75 36.72
Case-V	[60] 50.25	90 61.17	90 41.99	75 50.86
Case-VI	75 28.97	90 37.8	90 13.86	[60] 31.22

shown in Table 1. To maintain a fixed average utilization at each link in experiments, we place an additional (auxiliary) router (not shown in the figure) between node S_1 and router R_1, S_2 and R_1, S_3 and R_3, and S_4 and R_2 to limit the aggregate sending rate of the TCP flows to the capacity of the additional router. The utilization of R_i is controlled by properly setting the capacity of the auxiliary router.

Estimation Reliability. Using the above setup, we run Pathload with 4 different values of interrupt delays δ. To demonstrate estimation accuracy, we define the following relative error metric: $e_A = |A - \tilde{A}|/A$, where A is the true available bandwidth of a path and $\tilde{A}$ is its estimate. We report estimation results for each case in Table 2, which show relative estimation errors e_A of available bandwidth. As the table shows, with relatively small interrupt delays (e.g., $\delta \leq 100$ μs), Pathload estimates available bandwidth of the tight link with over 80% accuracy for all cases studied in this paper. Note, however, from the table that when δ becomes larger than 125 μs, it is unable to produce estimates for any of the cases as shown in the table as empty cells, which suggests that its algorithm is susceptible to non-trivial interrupt delays. We also conduct experiments with $\delta = 250$ and 500 μs and confirm its inability, but omit these results for brevity.

Table 2. Pathload's Measurement in Emulab

Interrupt	Evaluation scenario					
delay δ	Case-I	Case-II	Case-III	Case-IV	Case-V	Case-VI
0 μs	9.45%	8.00%	7.57%	6.48%	16.58%	15.01%
100 μs	1.44%	8.52%	14.9%	5.74%	3.6%	20.74%
125 μs	--	--	15.01%	--	--	34.65%
> 125 μs	--	--	--	--	--	--

Next, we investigate Pathload's internal algorithm in detail and identify what causes its measurement to be unstable under non-negligible values of δ.

3.2 Analysis

Recall that Pathload [9] sends back-to-back packets in a train of size $N = 100$ with a fixed rate R and examines one-way delay[2] (OWD) of each packet in the probe-train in order to identify a trend exists in the time-series delay data. Based on OWD delay trend, Pathload determines whether the current rate R is faster than the available bandwidth of the path under investigation. Hence, proper detection of OWD trend in a probe-train is crucial for it to produce an accurate and reliable bandwidth estimate of the path.

Note that Pathload first perform ADR (Asymptotic Dispersion Rate) probing by sending a single packet-train and checks interrupt moderation, which it detects when more than 60% of packets in a probe-train have been received back-to-back (with zero or negligible inter-packet delay). If interrupt moderation is detected, Pathload first eliminates such coalesced packets from the received train. Then, it directly performs PCT (Pairwise Comparison Test) and PDT (Pairwise Difference Test) on the remaining data if the number of remaining packets is no less than 5. Recall that the PCT metric represents the fraction of consecutive OWD pairs that are increasing, while the PDT metric quantifies how strong the difference between the first and last OWDs in the data set is. Define X_j to be the one-way delay of a packet j in a set of size n. Then, the PCT and PDT metrics[3] are given by [9]:

$$PCT = \frac{1}{n-1}\sum_{j=2}^{n} I(X_j > X_{j-1}), \quad PDT = (X_n - X_1)/\sum_{j=2}^{n}|X_j - X_{j-1}|, \quad (1)$$

where $I(Y)$ is one if Y holds, zero otherwise.

On the other hand, when Pathload does not detect interrupt moderation from the initial check, it first eliminates back-to-back packets from the probe-train just

---

[2] One-way delay of a packet is defined as the difference between its arrival time at the receiver and the corresponding sending time at the sender.

[3] Pathload [9] determines OWDs as "increasing" if $PCT > 0.66$, "non-increasing" if $PCT < 0.54$, or "ambiguous" otherwise. Similarly, it identifies OWDs as "increasing" if $PDT > 0.55$, "non-increasing" if $PDT < 0.45$, or "ambiguous" otherwise.

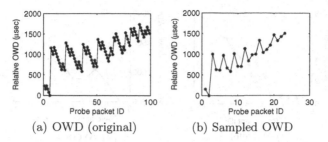

(a) OWD (original)   (b) Sampled OWD

Fig. 2. Relative OWDs obtained using the path in case I ($A = 31$ Mb/s)

like the previous case. If the number of remaining packets is no smaller than 36, then Pathload selects OWDs from the remaining packets using median-based sampling (see [9] for details) and applies the PCT and PDT tests to the sampled OWDs.

To assess Pathload's trend detection mechanism, we conduct experiments for Case I with interrupt delay $\delta = 250$ μs. In this example, we collect one-way delay data by running Pathload with a fixed rate $R = 38$ Mb/s and examine how its internal algorithm specifies a delay-trend existing in OWDs. Fig. 2(a) illustrates relative OWDs (one-way delays subtracted by their minimum value) obtained by sending packet trains at 38 Mb/s over the path in case I (available bandwidth $A = 31$ Mb/s). Note in the figure that OWDs exhibit an increasing trend over all even though they decrease in a small-scale burst (successive OWDs in the same burst decrease if the latency for transferring a packet from NIC to the user space at the receiver is smaller than the inter-packet dispersions exiting NIC at the sender [15]). Since the PCT and PDT tests cannot accurately detect a trend present in this kind of coalesced data, Pathload first removes coalesced packets before applying the PCT and PDT tests. Fig. 2(b) shows remaining OWDs after eliminating the coalesced packets. However, even with the data shown in Fig. 2(b), Pathload is unable to detect the increasing trend present in the data since its trend-test produces $PCT = 0.5$ and $PDT = 0.11$. This indicates that Pathload's trend-detection mechanism is not robust under the presence of coalesced packets due to interrupt delays.

Note that Pathload often discards entire packet-trains even with strong presence of a trend in the data due to its inability to detect the trend accurately. Although more extensive evaluations are required to confirm our findings, we believe that Pathload's inaccuracy in trend detection is the major problem that makes it unlikely to be successful in real networks.

4   IMR-Pathload

Motivated by the difficulty of characterizing delay variations in measured noisy OWD data, we study noise-filtering techniques such as wavelet-based signal processing and window-based averaging and explore their applicability in reliably identifying a trend from the data. In what follows below, we first investigate

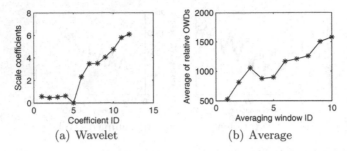

(a) Wavelet   (b) Average

Fig. 3. Wavelet coefficients and 10-packet window averages of relative OWDs shown in Fig. 2(a)

wavelet-based signal processing techniques that are widely used in removing noise from various data sets obtained empirically [2]. To overcome the effect of interrupt delays on trend detection, we apply a simple multi-level discrete wavelet transform [1] to OWDs before performing PCT- and PDT-based trend-test.

Note that in the multi-level wavelet decomposition, each stage consists of scale and wavelet filters followed by down-sampling by a factor of 2 and separates an input signal into two sets of coefficients: scale and wavelet coefficients. The wavelet coefficients represent a noise component in the input signal and thus are not processed further. On the other hand, the scale coefficients are applied to the two filters in the next level as an input to further reduce noise that might still exist in the scale coefficients from the previous stage. As a decomposition level increases, the frequency of wavelets used in filters decreases, capturing lower frequency components present in the original signal.

For experiments in this section, we use the family of Daubechies wavelets [3], which are well known standard wavelets (other wavelets can be used, but performance comparison among different wavelets is beyond the scope of this paper). Specifically, we use Daubechies' length-4 wavelets, whose scale filter coefficients are given by $h_0 = \frac{1+\sqrt{3}}{4\sqrt{2}}$, $h_1 = \frac{3+\sqrt{3}}{4\sqrt{2}}$, $h_2 = \frac{3-\sqrt{3}}{4\sqrt{2}}$, and $h_3 = \frac{1-\sqrt{3}}{4\sqrt{2}}$, while its wavelet filter coefficients are $g_0 = h_3$, $g_1 = -h_2$, $g_2 = h_1$, and $g_3 = -h_0$.

Assume that a sequence $s_0, s_1, \ldots, s_{n-1}$ is an input to the j-th stage filters. Define $cA_{j,k}$ and $cD_{j,k}$ (where $k = 0, 1, \ldots, n/2$) to be the scale and wavelet coefficients produced at level j, respectively. Then, $cA_{j,k}$ and $cD_{j,k}$ are given by:

$$cA_{j,k} = h_0 s_{2k} + h_1 s_{2k+1} + h_2 s_{2k+2} + h_3 s_{2k+3} \tag{2}$$

$$cD_{j,k} = g_0 s_{2k} + g_1 s_{2k+1} + g_2 s_{2k+2} + g_3 s_{2k+3}. \tag{3}$$

Note that when $k \geq n/2 - 1$, there are not enough data in the input sequence to compute the coefficients using (2) and (3). This is known as a boundary condition [17], which requires a special treatment that adds more data points to the input sequence (in this paper, we add the last value if necessary).

To demonstrate the effect of wavelet decomposition on trend detection, we decompose OWDs shown in Fig. 2(a) up to level 3 and plot in Fig. 3(a) the scale coefficients that represent the trend component of OWD data. Applying the

Table 3. Emulab Experiment

Estimation method	Interrupt delay δ	Evaluation scenario					
		Case-I	Case-II	Case-III	Case-IV	Case-V	Case-VI
IMR-Pathload (wavelet)	0 μs	2.46%	1.23%	3.47%	2.69%	3.71%	6.52%
	100 μs	6.47%	4.5%	3.02%	4.42%	5.98%	12.17%
	125 μs	7.21%	2.64%	3.88%	1.32%	6.1%	10.77%
	500 μs	5.12%	2.17%	6.78%	3.24%	7.23%	5.56%
IMR-Pathload (average)	0 μs	2.07%	2.24%	2.1%	2.18%	9.67%	5.05%
	100 μs	0.19%	0.71%	11.69%	1.32%	4.19%	6.82%
	125 μs	1.44%	1.82%	12.58%	1.59%	2.64%	7.89%
	500 μs	4.43%	4.59%	9.27%	2.55%	8.95%	6.48%

same PCT and PDT tests to the scale coefficient data, we get $PCT = 0.75$ and $PDT = 0.78$, which means that OWDs exhibit an increasing trend according to the criteria used in Pathload (recall that Pathload fails to detect this increasing trend as discussed in §3.2).

Next, we explore how window-based averaging improves trend detection in noisy data. In this approach, we take the average of OWDs in a window of size k (k-packet sliding window). Using a smaller window makes trend-detection susceptible to a larger interrupt delay since it may not sufficiently remove noise from OWDs (we leave optimal selection of window size as future work). For this example, we employ $k = 10$ and plot in Fig. 3(b) 10-packet window averages of relative OWDs shown in Fig. 2(a), which clearly shows an increasing trend. With these averaged OWDs, we get $PCT = 0.8$ and $PDT = 0.74$, which leads us to conclude that an increasing trend exists in the measured data.

We incorporate the above trend-detection mechanisms into Pathload and call it IMR-Pathload (*Interrupt Moderation Resilient Pathload*). We then evaluate it in Emulab and PlanetLab in the following section.

5   Performance Evaluation

5.1   Emulab Experiments

We investigate estimation accuracy of IMR-Pathload for different interrupt delays and report its relative estimation errors e_A in Table 3. As the table shows, IMR-Pathload produces available bandwidth estimates for all cases with $88-99\%$ accuracy, which is significantly better than that of Pathload (see Table 2). Notice in the table that even with a large interrupt delay $\delta = 500$ μs, IMR-Pathload measures the paths within $e_A = 10\%$ error in all studied cases (recall that Pathload can measure none of the paths if $\delta > 125$ μs as discussed in §3.1).

5.2   Internet Experiments

In this section, we report experimental results obtained by measuring several Internet paths between Universities and a HP Lab in U.S. Measurement hosts

Table 4. Internet Experiment

Internet paths	Method	Available bandwidth estimates (Mb/s)				
		9 – 10 am	12 – 1 pm	3 – 4 pm	7 – 8 pm	11 – 12 pm
HP → Wustl	IMR-Pathload	12.2	11.9	13	12.8	13.1
	Pathload	– –	– –	– –	– –	– –
UMD → HP	IMR-Pathload	93	92.8	92.3	93.2	94.7
	Pathload	95.1	91.7	91.2	93.2	92.6
UMD → TAMU	IMR-Pathload	100	98.1	98.3	99.4	98.4
	Pathload	– –	– –	– –	– –	– –
HP → UMD	IMR-Pathload	12.9	11.8	13.3	12.3	12.6
	Pathload	20	– –	16.9	– –	– –

used in this study are located at HP (HP Labs), TAMU (Texas A&M University), UMD (University of Maryland), and Wustl (Washington University in St Louis). Note that we choose these paths simply for the convenience of accessibility. Also note that the purpose of these experiments is not to compare estimation accuracy of bandwidth estimators since we do not know exact characteristics of these paths. Instead, we use this example to assess how reliably IMR-Pathload measures Internet paths compared to Pathload.

For this purpose, we select 5 different periods of time in a day and run IMR-Pathload and Pathload three times for each time period to measure a particular path. When a tool produces bandwidth estimates reliably in all three times for each period, we report their average as its available bandwidth estimate. If the tool cannot estimate bandwidth at least once in three trials, we consider that the tool is not able to measure that particular path reliably in that period. For IMR-Pathload, we test both wavelet- and averaging-based algorithms, but report only wavelet-based estimates since the other produces similar results.

Table 4 shows bandwidth estimates produced by IMR-Pathload and Pathload. As the table shows, IMR-Pathload reliably produces available bandwidth estimates for all studied paths in all measurement time periods. Note that for a path (UMD → HP), Pathload also produces estimates that are similar to those of IMR-Pathload[4]. However, Pathload is unable to reliably measure the other three paths (HP → Wustl, UMD → TAMU, and HP → UMD).

6   Conclusion

This paper studied Pathload under a wide range of end-host interrupt delays and identified its estimation instability under non-negligible interrupt delays. We found that Pathload's instability stems from that its delay-trend detection mechanism is unreliable when probing packets are coalesced at the receiver. We overcame this problem using robust trend detection algorithms (called

---

[4] This agrees with the Emulab results, where Pathload shows accuracy that is similar to IMR-Pathload only if it is able to reliably measure the path (see Tables 2 and 3).

IMR-Pathload) and showed using Emulab and Internet experiments that IMR-Pathload greatly improves measurement stability of Pathload under various network settings.

References

1. Burrus, C., Gopinath, R., Guo, H.: Introduction to Wavelets and Wavelet Transforms: A Primer. Prentice-Hall, Englewood Cliffs (1998)
2. Craigmile, P., Guttorp, P., Percival, D.: Trend Assessment in a Long Memory Dependence Model Using the Discrete Wavelet Transform. Environmetrics 15(4), 313–335 (2004)
3. Daubechies, I.: Orthonormal Bases of Compactly Supported Wavelets. Communications on Pure and Applied Mathematics 41(7), 909–996 (1988)
4. Dovrolis, C., Ramanathan, P., Moore, D.: Packet-Dispersion Techniques and a Capacity-Estimation Methodology. IEEE/ACM Trans. Netw. 12(6), 963–977 (2004)
5. Emulab. [Online], http://www.emulab.net/
6. Hu, N., Steenkiste, P.: Evaluation and Characterization of Available Bandwidth Probing Techniques. IEEE J. Sel. Areas Commun. 21(6), 879–974 (2003)
7. Interrupt Moderation Using Intel GbE Controllers. [Online], http://download.intel.com/design/network/applnots/ap450.pdf
8. Iperf – The TCP/UDP Bandwidth Measurement Tool. [Online], http://dast.nlanr.net/Projects/Iperf/
9. Jain, M., Dovrolis, C.: Pathload: A Measurement Tool for End-to-End Available Bandwidth. In: Proc. Passive and Active Measurement Workshop (March 2002)
10. Jin, G., Tierney, B.L.: System Capability Effects on Algorithms for Network Bandwidth Measurement. In: Proc. ACM IMC, October 2003, pp. 27–38 (2003)
11. Kang, S., Liu, X., Bhati, A., Loguinov, D.: On Estimating Tight-Link Bandwidth Characteristics over Multi-Hop Paths. In: Proc. IEEE ICDCS (July 2006)
12. Kang, S., Liu, X., Dai, M., Loguinov, D.: Packet-Pair Bandwidth Estimation: Stochastic Analysis of a Single Congested Node. In: Proc. IEEE ICNP, October 2004, pp. 316–325 (2004)
13. Kapoor, R., Chen, L., Lao, L., Gerla, M., Sanadidi, M.: CapProbe: A Simple and Accurate Capacity Estimation Technique. In: Proc. ACM SIGCOMM, August 2004, pp. 67–78 (2004)
14. Melander, B., Björkman, M., Gunningberg, P.: A New End-to-End Probing and Analysis Method for Estimating Bandwidth Bottlenecks. In: Proc. IEEE GLOBECOM, November 2000, pp. 415–420 (2000)
15. Prasad, R., Jain, M., Dovrolis, C.: Effects of Interrupt Coalescence on Network Measurements. In: Proc. Passive and Active Measurement Workshop (April 2004)
16. Ribeiro, V., Riedi, R., Baraniuk, R., Navratil, J., Cottrell, L.: pathChirp: Efficient Available Bandwidth Estimation for Network Paths. In: Proc. Passive and Active Measurement Workshop (April 2003)
17. Strang, G., Nguyen, T.: Wavelets and Filter Banks. Wellesley-Cambridge Press (1996)

A Measurement Study of Internet Delay Asymmetry

Abhinav Pathak[1], Himabindu Pucha[1], Ying Zhang[2], Y. Charlie Hu[1],
and Z. Morley Mao[2]

[1] Purdue University
[2] University of Michigan
pathaka@purdue.edu, hpucha@purdue.edu, wingying@umich.edu,
ychu@purdue.edu, zmao@umich.edu

Abstract. RTT has been widely used as a metric for peer/server selection. However, many applications involving closest peer/server selection such as streaming, tree-based multicast services and other UDP and TCP based services would benefit more from knowing one-way delay (OWD) rather than RTT. In fact, RTT is frequently used as as an approximate solution to infer forward and reverse delays by many protocols and applications which assume forward and reverse delay to be equal to half of RTT.

In this paper, we compare and contrast one-way delays and corresponding RTTs using a wide selection of routes in the Internet. We first measure the extent and severeness of asymmetry in forward and reverse OWD in the Internet. We then attempt to isolate the causes of OWD asymmetry by correlating OWD asymmetry with the route asymmetry. Finally, we investigate the dynamics of delay asymmetry. We find there exists a weak correlation between the fluctuation of RTT and OWD but a strong correlation between OWD change and the corresponding route change.

1 Introduction

Today's Internet is rife with several wide-area network applications: real-time applications such as voice over IP [1] and multicast streaming applications [2,3,4,5], data transfer applications that perform locality-aware redirection and server selection [6], and services such as proximity-aware DHTs [7,8] and positioning systems [9,10,11]. A common thread in all these applications is the requirement to perform proximity measurements. For example, in multicast applications, proximity is used to choose a suitable parent/child in the tree; in positioning systems, proximity to landmarks is used for localization.

For some applications, the proximity of interests can be measured using the round-trip time (RTT) between two end hosts (say A and B), defined as the sum of forward delay from A to B and the reverse delay from B to A. For example, if the interaction between remote hosts typically involves only one or a few request and reply messages, for example, a DNS lookup, or a small HTTP document download, then RTT is a good indication of the completion time of the interaction. For other applications, however, the proximity of direct relevance is the one-way delay (OWD) from the client to the servers/peers or along the other direction, rather than RTT. An asymmetry in OWD could hurt such applications. For example, in multicast streaming applications, since

M. Claypool and S. Uhlig (Eds.): PAM 2008, LNCS 4979, pp. 182–191, 2008.

the data always flows from a parent node to a child node in the overlay multicast tree, optimizing the OWD from the parent to the child is more beneficial. As another example, ACK/NAK data systems such as the Transmission Control Protocol (TCP) estimate the available bandwidth of the unidirectional route from the sender to the receiver using the round-trip time as an approximation. If the reverse path taken by ACK packets has a much larger delay than the forward path delay, TCP can end up using more network resources than it should [12]. Other applications that depends on OWD include online multiplayer games where an asymmetry in delay could create a bias, video conferencing applications, Internet distance prediction, etc.

In practice, measuring the OWD between two end hosts, however, faces two major obstacles. First, it requires strict time synchronization between the two hosts. Second, it requires access to both end hosts as there is no standard daemon in operating systems that measures and reports OWD. As a result, the standard practice in almost all applications that rely on proximity information is to measure RTT and operate under the assumption that OWD is half of RTT.

In this work, we investigate the validity and implications of this premise (and common practice) via a comprehensive measurement study: We first measure the extent and severeness of asymmetry in forward and reverse OWD between several pairs of Internet hosts. We then attempt to trace the reasons for the observed OWD-RTT relationship. Further, we investigate the dynamics of the OWD-RTT relationship as both OWD and RTT change. Our main findings are as follows: (1) Asymmetry between the forward and reverse delays is quite prevalent. (2) Asymmetry in delay can be attributed at least in part to the asymmetry in routing paths. (3) Delay asymmetry is dynamic - with progression of time, delay asymmetry varies. To track how the asymmetry varies as RTT and route changes happen, we make the following observations: During an intra-AS path change, in most cases the forward and reverse delay change equally. During an inter-AS path change, either both forward and reverse delay change equally (keeping asymmetry constant), or only forward delay change contributes to RTT change (changing asymmetry).

2 Methodology

2.1 Tools, Testbed and Trace Collection

We use owping [13] for measuring OWD between a source-destination pair. Owping is an implementation of One-way Active Measurement Protocol (OWAMP) [14]. It requires access to both end hosts between whom the delay is to be measured. The destination node runs one-way ping server, "owampd" (owamp daemon), which listens for client requests for conducting one-way ping measurements. The source node initiates the measurement using owping (client). For every measurement, the source and destination exchange 10 probe packets.

Owping requires time synchronization between the end hosts running the measurements. OWD between Internet hosts range from a few milliseconds to hundreds of milliseconds. Even a minimal clock drift could result in inaccuracy in OWD measurements. To capture clock drifts (difference between node's clock and NTP's clock), before reporting any "time" measurements, the owping tool gets the current clock drift of the local node from the NTP daemon running on the system. The clock drift is adjusted in the time reported by the system. NTP daemon also reports error estimate (confidence value) of the clock. Error estimate gives the confidence range (in milliseconds) of time reported by NTP. The source marks its current time (adjusted with clock drift) on the packet along with its current error estimate with respect to NTP before sending the packet. Upon receiving the packet, the destination subtracts the time on the packet from its current time (fetched using kernel timestamp of the packet to minimize error due to load on PlanetLab nodes and then corrected with its clock drift). It also notes down error estimate at its end due to NTP. The overall error estimate for a single probe packet is calculated by adding up the corresponding error estimates of both ends. For all the 10 packets, the minimum, mean and maximum values of the 10 OWD are recorded, along with the maximum error estimate of the 10 error estimates. To measure route between a pair of nodes, we used Paris traceroute [15].

Since owping requires access to both end hosts for which the delay is measured, we use the PlanetLab [16] testbed for our measurement study. PlanetLab contains nodes belonging to both research/education networks (GREN) and commercial networks. A recent study [17] has shown that the network properties of the paths between two hosts in GREN (denoted as G2G - Gren-to-Gren) can be very different from those in the commercial networks. The same study also showed that if at least one end of the path is in the commercial network (G2C or C2G), then the network properties remain similar compared to when both ends are in the commercial network (C2C). Hence, we perform this study by separating the network paths into all GREN and commercial paths (G2C+C2G+C2C). We chose 180 GREN nodes and 25 commercial nodes.

We collected traces from April 12th, 2007 for a period of 10 days. Traces consist of back-to-back traceroutes and OWD measurements. One measurement round consists of a traceroute and a OWD measurement to every other node in our set and was repeated after every 20 minutes. The data collected was stored on a central node for parsing and further processing. We continuously monitored over 10000 Internet paths during this period. We conducted over 5 million traceroutes and an equal number of OWD estimations during the specified period. We minimized the error induced due to loaded PlanetLab nodes [18] by taking the minimum of 10 readings reported by owping.

2.2   Trace Pruning

Accurate one-way ping measurements critically depends on end nodes synchronization. When a host reboots, its clock is not synchronized and hence its measurements are not usable. However, after booting up, it contacts an NTP server and is configured to correct its clock [19]. This adjustment consumes some time after which the node is synchronized with NTP. In general, the uptime of PlanetLab nodes is large. However, when a node goes down, we remove all the measurements to and from it. We also eliminate nodes with large error estimates. Figure 1 plots the CDF of clock drift and the maximum error estimate for PlanetLab nodes. We observe that PlanetLab nodes do not drift much from their NTP servers. 60% of the nodes drift less than 2 milliseconds. We also observe that the maximum error estimated by NTP is lower than 10 ms for about 40% of the nodes. As a first cut we chose 10ms error estimate threshold and pruned

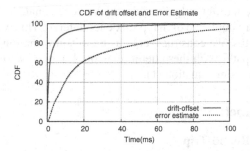

Fig. 1. CDF of drift offset and error estimate among 205 nodes in PlanetLab

our node selection by eliminating nodes that deviated more. After this filtering, we had 82 GREN nodes and 12 commercial nodes. Further, when computing any ratios using absolute forward and reverse delays, we limit the values of error estimate. For example, in Section 3 wherever we took ratios involving OWD, such as forward delay by RTT, we made sure that the error estimated in forward delay stays below 3% of the measured RTT. In Section 4, we deal with change in forward and reverse delays at a node, and correlate them to change of RTT. Whenever we take a difference, of say forward delay, the error in the value gets nullified. This way we do not need additional data pruning.

2.3   Metrics

We ultimately relate delay asymmetry on the Internet to other asymmetric properties such as network path. In this section we define metrics that we used to quantify the level of path asymmetry among the measured routes. Our metric needs to adequately capture the properties of an Internet path responsible for inducing delays in the network. The packet delay introduced by the path depends on the intermediate routers and ASes through which the packet travels. We compare two metrics to characterize these properties - AS-level path asymmetry and router-level path asymmetry. **AS-level path asymmetry** captures the dissimilarity among ASes in the forward and the reverse paths. We quantify AS-level path asymmetry using AS-level path similarity coefficient (γ) as follows: For every source destination pair, let the set of ASes in the forward path (P_f) and the reverse path (P_r) be A and B respectively. The similarity coefficient γ_{P_f,P_r} is calculated as

$$\gamma_{P_f,P_r} = \frac{|A \cap B|}{|A \cup B|} \tag{1}$$

Router-level path asymmetry captures the dissimilarity between the forward and the reverse path at the IP-level hops. To determine the router-level asymmetry in the forward and reverse path, we can not simply use IP addresses in the traceroutes in the forward and reverse directions because many routers have different interfaces to handle traffic in different directions. Most of these interfaces lie in the same /24 prefix. Grouping IP addresses of intermediate routers obtained with same /24 prefixes into one loses the hop count information of the path. To avoid these problems we use the following approach: In each path we took the /24 prefixes of the IP addresses of the intermediate routers. If there exists more than one interface belonging to the same /24 prefix in the path, we counted them differently. Specifically we concatenated an increasing counter per /24 prefix to every duplicate of that prefix encountered. This was done for both forward and reverse paths. We construct set A using this information in forward path and set B from reverse path. With these definitions of A and B, we compute router-level similarity coefficient using equation (1).

3   Delay Asymmetry

3.1   One-Way vs. Round-Trip

OWD in forward and reverse directions add up to give RTT, with a general perception that the forward and reverse delays are equal. This section investigates the extent and the severeness of delay asymmetry among GREN and commercial paths individually. To that end, we observe the correlation between OWD and RTT, as shown in Figure 2.

Figure 2(a) shows that paths in G2G are largely symmetric with respect to their delays (CDF remaining close to 0.5 shows symmetry). On the other hand, Figure 2(b) indicates that asymmetry is prevalent in commercial networks. In fact, the magnitude of asymmetry (the ratio of OWD in forward direction to RTT) varies from values below 0.4 to those above 0.6. Figure 2(c) augments 2(b) and shows the absolute values of forward delay and RTT for non-G2G paths. Figure 2(c) shows the magnitude of asymmetry. For example, we take the point where RTT is 150ms and forward delay is 60ms. This case would correspond to 40% of forward delay ratio. The reverse path delay accounts for 90ms. This results in 30ms of delay asymmetry in the forward and reverse paths. We see that asymmetry in delay indeed exists in today's Internet especially in commercial networks.

3.2   Asymmetry in One-Way Delay

We now dig a deeper to understand the origin of this asymmetry in delay values. Intuitively, if the forward and the reverse paths between a source-destination pair are

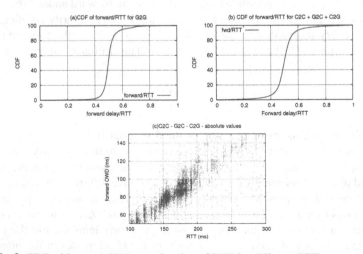

Fig. 2. CDF of forward OWD as a fraction of RTT for different RTTs occurrences

Table 1. Various possibilities observed that change delay asymmetry

Legend	RTT Changes	Forward Changes	Reverse Changes	Explanation
A	Yes	No	Yes	RTT changes due to a change in reverse delay but forward delay is unchanged
B	Yes	Yes	No	RTT changes due to change in forward delay but reverse delay is unchanged
C	Yes	Yes	Yes	RTT changes along with changes in forward and reverse delays
D	No	Yes	Yes	RTT does not change but both forward and reverse delay changes (equally and opposite in sign)

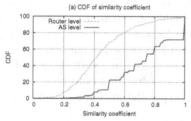

(a) CDF of AS-level and router-level similarity coefficient in forward and reverse path of all routes observed.

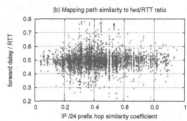

(b) Scatter plot of router-level similarity coefficient for each route vs. ratio of forward delay and RTT.

Fig. 3. Comparing path asymmetry with delay asymmetry

different, we can expect the corresponding properties to vary too. Path asymmetry is a well-known fact prevalent in the Internet. Many previous studies have found that a large amount of asymmetry exists in Internet routes [20,21,22]. In this section we study whether there exists any correlation between route asymmetry and delay asymmetry.

Figure 3(a) plots the CDF of AS-level and router-level similarity coefficients (defined in Section 2.3) for all the possible routes that were found in our trace. Figure 3(a) shows that 32% of paths have AS similarity coefficient of less than 0.6, while 81% of routes have router-level asymmetry coefficient of less than 0.6. The graph shows, as expected, that router-level asymmetry is more prevalent than AS-level asymmetry. Hence we use router-level similarity coefficient to characterize path asymmetry.

To find correlation between delay and path asymmetry we plot delay fraction, the ratio of forward delay by RTT, as a function of router-level similarity coefficient for each route observed. Figure 3(b) shows the correlation. From the figure we see that the delay fraction remains close to 0.5 when the router-level similarity coefficient is close to unity. In such cases, the forward and reverse paths traverse nearly the same set of routers and experience equal delays. They contribute equally to RTT. In cases when the router-level similarity coefficient is not close to unity the delay fraction fluctuates from 0.3 to 0.7. This gives us an indication that if there exists a significant router-level asymmetry, the forward and reverse OWD could be significantly different. In summary router-level asymmetry does not necessarily imply delay asymmetry where as delay asymmetry implies router-level asymmetry.

4   Dynamics of Delay Asymmetry

Although existence of delay asymmetry is of interest, another intriguing question is: Is the delay asymmetry for a given source-destination pair constant across time? If delay asymmetry remains constant then we could do a one-time measurement and tune applications accordingly. If not, how does it vary? Table 1 categorizes the various possibilities that can change delay asymmetry. From the table we see four reasons can cause delay asymmetry change.

To get a better picture of the prevalence of the above scenarios, we log the fluctuations in RTT values across our trace: Whenever RTT fluctuated by 2% for any source/destination pair, we note down the corresponding fluctuation in forward and reverse delays. Figure 4(a) plots a scatter plot to show the correlation between forward OWD change and RTT fluctuation. We see that the plot can be broken down into 4 major regions. The first region is parallel and close to x-axis (line y=0) (legend A of table 1), the second region along y=x line (legend B of table 1), a third region along y=0.5x line (legend C of table 1), and the fourth region parallel and close to y axis (x=0 line) (legend D of Table 1).

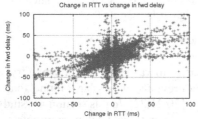

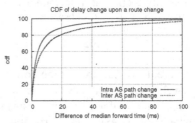

(a) RTT change vs forward OWD upon a 2% change in RTT.

(b) CDF of forward delay change in upon an inter-/intra-AS path change

Fig. 4. Dynamics of delay asymmetry

Thus, there is conclusive evidence that delay asymmetry is a dynamic property. Delay asymmetry changes when delays change. [23] points out two reasons for delay changes - path change and transient congestion. Out of these two major reasons of delay changes, we focus on delay changes caused due to a path change in the forward direction. These path changes are measured using periodic traceroutes from sources to destinations. Using repeated traceroutes we find out occurrences of a route change in forward direction (at a granularity of 20 minutes). At the same time, we traceroute the path in the reverse direction and also measure one-way delays. We divide the path change in forward direction into two categories, inter-AS path change and intra-AS path change.

Figure 4(b) shows by how much forward delay changes upon a change in the forward path. For every path change we observed in our measurement, we classified the path

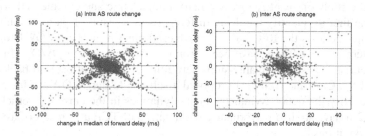

Fig. 5. Scatter plot: Reverse vs forward delay change for Intra-/Inter-AS path changes

change into inter-AS path change if the new path has a different AS-level path then the previous path, or intra-AS path change if there is a path change internal to an AS. We see that 80% of intra-AS path change results in change in forward delay by less than 10 milliseconds. Upon an inter-AS path change, about 80% of times the forward delay changes by less than 20 milliseconds.

A path change might change the delay asymmetry for a route. To find out the impact of path changes on change in OWD and RTT, for every inter and intra-AS path change that we observed, we plot a scatter plot between the absolute change in forward delay and absolute change in reverse delay. Figure 5(a) shows the correlation between forward and reverse delay changes upon a forward intra-AS path change. We see that there is a strong correlation between the attributed delay change values. In most of the cases, the change in forward and reverse delay equally contribute to the change in RTT (the patch of dots along the y=x line). This implies that when we notice a path change in the forward direction, in most of the cases the path in the reverse direction also changes.

Figure 5(b) shows correlation between forward and reverse delays when there is an inter-AS forward path change. There can be two cases here: the AS change takes place only in the forward direction, and the AS change in the forward path also affects the reverse path. Inter-AS path changes are less frequent than intra-AS path changes. The scatter plot in figure 5(b) shows two faint regions, one along the line y=x and the second region parallel and close to x-axis. The first region signifies of AS path change when forward and reverse delays equally contribute to RTT. This region captures the situation when forward AS path change also affects path in the reverse direction. The second region depicts the situation when AS-level path changes in the forward direction, forward delay is changed but reverse delay is not affected. In this case, the change in RTT is contributed solely by forward delay.

In summary, we have identified causes for delay symmetry changes based on routing changes. The property of delay asymmetry is found to be a dynamic property that varies depending on routing dynamics. As expected, inter-AS path changes contribute to larger delay changes compared to intra-AS path changes. Moreover, intra-AS path changes tend to have similar effect in terms of delay changes for both forward and reverse paths.

5 Related Work

Paxson [20] was among the first to study asymmetry in Internet routes. He et. al. [22] quantified the level of path asymmetry that exists in the current Internet. They conducted a systematic study of quantifying asymmetry in academic and commercial nodes. Our work is the first to quantify the delay differences as a result of path asymmetry. Our measurement follows the RFC 2679 [24] which defines a metric for OWD of packets across Internet paths. There have been proposals for measuring OWD using mathematical heuristics based on relationships between a sequence of back-to-back one-way packet delays. Choi and Yoo [25] proposed a scheme to derive OWD. However, they did not quantify the delay asymmetry or its dynamics in the current Internet. Several work [12,26] have studied performance degradation of TCP as a result of asymmetric network conditions.

6   Conclusion

In this paper, we studied delay asymmetry that exists in the current Internet. We found that commercial networks exhibit higher levels of asymmetry than education and research networks. We found a weak correlation between router-level path asymmetry and delay asymmetry. We then studied how delay asymmetry changes over time as path change occurs. We found that any of the four scenarios could take place (a) forward delay change equals RTT change with no change in reverse delay, (b) forward and reverse delays change equally contribute to a change in RTT, (c) reverse delay change contributes to entire RTT change with negligible forward delay change, and (d) forward delay and reverse delay change equally but in opposite directions, resulting in no effect on RTT. We also correlated properties of delay changes upon an inter- and intra-AS route change. Our work provides important foundations to enable applications to predict OWD changes as a result of possible routing changes and more accurately infer OWD values from RTT measurements.

Our findings suggest that proximity-based applications that need information about OWD can benefit from accurate OWD measurement as opposed to using half of RTT as an approximation. However, OWD measurement requires the cooperation of both end hosts. One possible solution to this is to incorporate OWD measurement software as a daemon in commodity OSes similarly as the ICMP echo daemon for normal ping.

References

1. Skype: The whole world can talk for free, http://www.skype.com/
2. Chu, Y.H., Rao, S.G., Zhang, H.: A case for end system multicast. In: Proc. of ACM SIGMETRICS (2000)
3. Zhang, B., Jamin, S., Zhang, L.: Host Multicast: A Framework for Delivering Multicast To End Users. In: Proc. of IEEE INFOCOM (June 2002)
4. Banerjee, S., Bhattacharjee, B., Kommareddy, C.: Scalable application layer multicast. In: Proc. of ACM SIGCOMM (2002)
5. Castro, M., Druschel, P., Kermarrec, A.M., Nandi, A., Rowstron, A., Singh, A.: SplitStream: High-Bandwidth Multicast in Cooperative Environments. In: Proc. of ACM SOSP (2003)
6. Akamai: Expertise content delivery, http://www.akamai.com/
7. Rowstron, A., Druschel, P.: Pastry: Scalable, Distributed Object Location and Routing for Large-Scale Peer-to-peer Systems. In: Proc. of Middleware (2001)
8. Zhao, B.Y., Huang, L., Stribling, J., Rhea, S.C., Joseph, A.D., Kubiatowicz, J.: Tapestry: A Resilient Global-Scale Overlay for Service Deployment. IEEE JSAC (2004)

9. Ng, T.S.E., Zhang, H.: Predicting Internet Network Distance with Coordinates-Based Approaches. In: Proceedings of IEEE INFOCOM (June 2002)
10. Dabek, F., Cox, R., Kaashoek, F., Morris, R.: Vivaldi: A Decentralized Network Coordinate System. In: Proceedings of ACM SIGCOMM (August 2004)
11. Francis, P., et al.: An Architecture for a Global Internet Host Distance Estimation Service. In: Proceedings of IEEE INFOCOM (March 1999)
12. Balakrishnan, H., Padmanabhan, V.N., Katz, R.H.: The effects of asymmetry on tcp performance. In: Proc. of ACM MobiCom (September 1997)
13. Owping: One way ping, http://e2epi.internet2.edu/owamp/
14. Shalunov, S., Teitelbaum, B., Karp, A., Boote, J., Zekauskas, M.: A One-way Active Measurement Protocol (OWAMP). In: RFC 4656 (Proposed Standard) (September 2006)
15. Augustin, B., et al.: Avoiding traceroute anomalies with Paris traceroute. In: Proc. of IMC (2006)
16. PlanetLab: An open platform for developing, deploying and accessing planetary scale services, http://www.planet-lab.org/
17. Pucha, H., Hu, Y.C., Mao, Z.M.: On the Representativeness of Wide Area Internet Testbed Experiments. In: Proc. of ACM IMC (2006)
18. Sommers, J., Barford, P.: An Active Measurement System for Shared Environments. In: Proceedings of IMC (October 2007)
19. Mills, D.L.: RFC 1305: Network time protocol (version 3) specification, implementation. Obsoletes RFC0958, RFC1059, RFC1119 Status: DRAFT STANDARD (March 1992)
20. Paxson, V.: End-to-end routing behavior in the Internet. In: Proc. of ACM SIGCOMM (1996)
21. Allman, M., Paxson, V.: On estimating end-to-end network path properties. In: Proc. of SIGCOMM (1999)
22. He, Y., Faloutsos, M., Krishnamurthy, S., Huffaker, B.: On routing asymmetry in the internet. In: Proceedings of IEEE Globecom 2005 (2005)
23. Pucha, H., Zhang, Y., Mao, Z.M., Hu, Y.C.: Understanding Network Delay Changes Caused by Routing Events. In: Proc. of ACM SIGMETRICS (2007)
24. Almes, G., Kalidindi, S., Zekauskas, M.: A One-way Delay Metric for IPPM. In: RFC 2679 (September 1999)
25. Choi, J.H., Yoo, C.: One-way delay estimation and its application. Computer Communications 28(7), 819–828 (2005)
26. Balakrishnan, H., Padmanabhan, V., Fairhurst, G., Sooriyabandara, M.: TCP performance implications of network path asymmetry. In: IETF RFC 3449 (2002)

Trends and Differences in Connection-Behavior within Classes of Internet Backbone Traffic

Wolfgang John, Sven Tafvelin, and Tomas Olovsson

Department of Computer Science and Engineering
Chalmers University of Technology, Göteborg, Sweden
{johnwolf,tafvelin,tomas}@chalmers.se

Abstract. In order to reveal the influence of different traffic classes on the Internet, backbone traffic was collected within an eight month period on backbone links of the Swedish University Network (SUNET). The collected data was then classified according to network application. In this study, three traffic classes (P2P, Web and malicious) are compared in terms of traffic volumes and signaling behavior. Furthermore, longitudinal trends and diurnal differences are highlighted. It is shown that traffic volumes are increasing considerably, with P2P-traffic clearly dominating. In contrast, the amount of malicious and attack traffic remains constant, even not exhibiting diurnal patterns. Next, P2P and Web traffic are shown to differ significantly in connection establishment and termination behavior. Finally, an analysis of TCP option usage revealed that Selective Acknowledgment (SACK), even though deployed by most web-clients, is still neglected by a number of popular web-servers.[1]

1 Introduction

Today, many network operators do not know which type of traffic they are carrying. This problem emerged mainly in the early 2000's, when P2P file sharing applications started to disguise their traffic in order to evade traffic filters and legal implications. Since then, the network research community started to draw increasing attention to classification of Internet traffic. Traditional port number classification was shown to underestimate actual P2P traffic volumes by factors of 2-3 [1], thus more sophisticated classification methods have been proposed. These methods are typically either based on payload signatures [2], statistical properties of flows [3] or connection patterns [4].

A number of articles also present properties of different traffic classes resulting from traffic classification. Gerber et al. [5] classified flow measurements from a tier-1 ISP backbone in 2003. Even if their classification method has been based on port numbers, they indicate a dominance of P2P applications. Sen et al. [6] investigated connectivity aspects of P2P traffic on different levels of aggregation (IP, prefix, AS) in 2002. The study was based on flow data collected at a single ISP, classified by a port number method. More recent articles from 2005 and 2006

---

[1] This work was supported by SUNET, the Swedish University Network.

M. Claypool and S. Uhlig (Eds.): PAM 2008, LNCS 4979, pp. 192–201, 2008.

present differences between P2P and non-P2P traffic in terms of flow properties such as size, duration and inter-arrival times [7,8]. Perenyi et al. [8] additionally presents a comparison of diurnal patterns for P2P vs. non-P2P traffic.

This article presents the results of a classification of current Internet backbone data. The datasets do not include packet payloads, thus connection pattern heuristics [9] were used to classify the datasets. The classification approach, disregarding packet payload data, has the advantage of avoiding legal issues and has the capability to classify even encrypted traffic, which is gaining popularity among P2P traffic. We chose to focus on 3 main traffic classes: (1) P2P file sharing protocols; (2) Web traffic; (3) malicious and attack traffic. First, we show how these traffic classes develop over a time period of eight months by highlighting trends in traffic volumes and connection numbers, also pointing out some diurnal differences. Next, we present differences between the traffic classes in terms of connection signaling behavior. This includes success rates for TCP connection establishment, a breakdown of different TCP connection termination possibilities and TCP option usage within established connections.

To our knowledge, this is the first attempt to characterize differences and trends within traffic classes in terms of connection signaling, with exception of a brief discussion about connection termination in [10]. We provide a thorough analysis of differences and trends for the selected traffic classes, since they have a major impact on the overall traffic behavior on the Internet. It is of general importance to follow trends in contemporary Internet traffic in order to react accordingly in both infrastructure and protocol development. Furthermore a thorough analysis of specific connection properties reveals how different traffic classes are behaving 'in the wild'. Since the data analyzed was collected on a highly aggregated backbone during a substantial time period, the results reflect contemporary traffic behavior of one part of the Internet. These results are thereby not only valuable input for simulation models, they are also interesting for developers of network infrastructure, applications and protocols.

2   Data Description

The two datasets used in this article [11] were collected in April (spring dataset) and in the time from September to November 2006 (fall dataset) on an OC192 backbone link of the Swedish University Network (SUNET). In spring, four traces of 20 minutes were collected each day at identical times (2AM, 10AM, 2PM, 8PM) as described in [12]. The fall dataset was collected at 276 randomized times during 80 days. At each random time, a trace of 10 minutes duration was stored. To avoid bias when comparing the datasets, the 20 minute samples from spring were treated as two separate 10 minute traces. Furthermore, for this study traces from fall are only considered if collected during the time-window between 20 minutes prior and after the collection times of spring (e.g. 1:40AM-2:40AM).

When recording the packet level traces on the 2x10GB links, payload beyond transport layer was removed and IP addresses were anonymized due to privacy concerns. After further pre-processing of the traces, as described in [11] and

[12], a per-flow analysis was conducted on the resulting bi-directional traces. Flows are defined by the 5-tuple of source and destination IP, port numbers and transport protocol (TCP or UDP). TCP flows represent connections, and are therefore further separated by SYN, FIN and RST packets. For UDP flows, a flow timeout of 64 seconds was used [4]. The 146 traces in the spring dataset include 81 million TCP connections and 91 million UDP flows, carrying a total of 7.5 TB of data. The reduced fall dataset, consisting of 65 traces, includes 49 million TCP connections and 70 million UDP flows, carrying 5 TB of data. In both datasets, TCP connections are responsible for 96% of all data.

3   Methodology

The resulting 130 million TCP connections and 161 million UDP flows have been fed into a database, including per-flow information about packet numbers, data volumes, timing, TCP flags and TCP options. The flows have then been classified by use of a set of heuristics based on connection patterns. The classification method was introduced and verified on the April dataset, as described in [9]. The heuristics are intended to provide a relatively fast and simple method to classify traffic, which was shown to work well on traces even as short as 10 minutes. In the present study the flows are summarized into three different traffic classes: P2P (file-sharing); Web or HTTP (incl. HTTPS); Malicious and attack (i.e. scan, sweep and DoS attacks). Remaining traffic was binned in a fourth class, denoted 'others'. 'Others' includes mail, messenger, ftp, gaming, dns, ntp and remaining unclassified traffic. The latter accounts for about 1% of all connections. In this study, the focus is on trends and differences between P2P and Web traffic, with some notable observations from malicious traffic highlighted as well. Besides the traffic classification, an analysis of traffic volumes and signaling properties is carried out in two further dimensions: longitudinal trends between April and November and diurnal patterns between the four time clusters (times of day).

4   Trends in Traffic Volumes

Longitudinal trends in TCP traffic volumes have been analyzed by building time series for the three traffic classes within each of the four time clusters, representing times of day (2AM, 10AM, 2PM, 8PM). Due to space limitations, only a condensed time series of TCP traffic is illustrated in Fig.1. The x-axis of the graphs represent time, with one bar for each 10 minute long trace. The first row indicates an increase in traffic volume during 2006. While peak volume per 10 minutes lies at 70 GB in early April, volume reaches 85 GB in late April (right after Easter vacation). This trend continues, with peaks of 94 GB in September and finally 113 GB in November. During one specific interval on November 8 as much as 131 GB have been transfered via TCP. All peak intervals fall into the time cluster of 8PM. The second busiest time cluster in terms of traffic volumes is the one at 2PM. Transfer volumes during 2PM reach on average 80% of the peak values at 8PM. Nighttime and morning hours (2AM, 10AM) show the

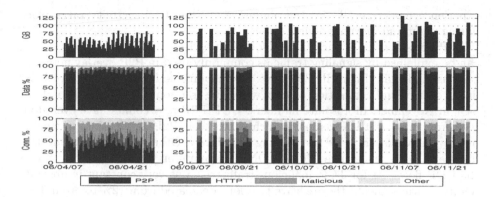

Fig. 1. TCP data vs time (1st row); Appl. breakdown by data(2nd) and #conn.(3rd)

lowest activity with half the transfer volumes of the busy evening hours. This diurnal pattern is best visible in the April section of the first row in Fig.1.

Even if there is an increase in data volumes of around 65% during a time period of eight months, the breakdown into traffic classes remains constant. P2P applications account constantly for as much as 93% and 91% of the data during evening and night time, respectively. During office hours (10AM, 2PM) the fraction of P2P data is reduced to 86%. HTTP, in contrast, is responsible for 9% of TCP data transfered during office hours, and drops down to 5% and 4% during evening and night time. This diurnal difference is explained by a network prefix analysis, yielding that most P2P traffic originates from student dormitories whereas Web traffic is commonly generated by Universities. The remaining data fractions account mainly for 'other' traffic, since malicious traffic and attacks tend to be single packet flows, not carrying substantial amounts of data.

The traffic breakdown in terms of connection numbers clearly shows that P2P connections typically carry higher amounts of data. Between 40% and 55% of the connections are classified as P2P, following the diurnal patterns of traffic volumes. HTTP connections account for 25% of all TCP connections during office hours, but drop down to 7% at night hours. Interestingly, the fractions of both P2P and HTTP connections (or connection attempts) increased slightly from April to November, while the fraction of malicious traffic decreased from around 30% to 20% during the same time. This development turns out to be a consequence of the constant nature of malicious traffic, such as scanning attacks. In absolute numbers, this traffic class remained remarkably constant during the eight months. Due to the increase in overall traffic volume, its relative fraction evidently was decreased. Since malicious or attack traffic shows neither longitudinal trends nor any significant diurnal pattern, we conclude that this type of traffic rather forms a constant 'background noise' in the Internet.

A similar analysis was also done for UDP flows. Even though larger in number, they are only responsible for 4% of all data. UDP data volumes during 10 minutes increased from peak values of 2.8 GB in April up to 4.6 GB in November. As in the case of TCP, peak intervals fall into the 8PM time cluster. Afternoon hours experience moderate UDP data volumes, and little UDP activity takes place during night and morning hours.

P2P flows over UDP carry in 76% of all cases less than three packets, which can be explained by signaling traffic as commonly used in P2P overlay networks such as Kademlia. In April, P2P flows are responsible for around 80% of UDP data volumes and connection counts, while the fraction has increased to about 84% in November. In absolute numbers, UDP P2P flow counts have even doubled from April until November, which shows that P2P applications deploying overlay networks via UDP are gaining popularity. Other traffic, including traditional UDP services like NTP or DNS, accounts on average for only 8% of the UDP flows. As for TCP, malicious traffic remains very constant in absolute numbers, which means that relative fractions decreased from 12% to around 8% in November.

5   Differences between Traffic Classes

The following subsection highlights differences between P2P, Web and malicious connections in terms of establishment and termination behavior. In the next subsection, TCP option deployment for P2P and Web connections is compared.

5.1   Differences in Connection Behavior

Fig.2 breaks down the success-rates of connection attempts for the three classes. Established connections include TCP flows with successfully carried out 3-way-handshakes. The second group of connection attempts did not fulfill 3-way-handshakes, but included an initial SYN packet. Finally, there are flows with no SYN seen. These are TCP sessions starting before the measurement interval. Such session fragments account for 13.5% of the 130 million connections seen. Malicious traffic usually consists of 1-packet flows only, which explains why only few malicious connection attempts fall into the no SYN category. In the further analysis, we will only focus on connections including initial SYN packets.

A notable trend can be observed in the P2P graph in Fig.2, where the fraction of unsuccessful connection attempts increased from an average of 49% in April to 54% in November. Web traffic on the other hand has significantly larger fractions

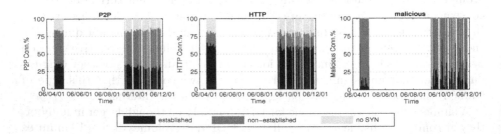

Fig. 2. TCP Connection Breakdown

of established connections, leaving only an average of 16.3% non-established. Malicious traffic is more likely to be established in the fall data, even though a majority of the malicious connections are still connection attempts. The increase in established attack connections is caused by an increase in login attempts to MS-SQL and SSH servers, with a few MS-SQL servers at a local University responsible for the majority of the attempts. According to SANS Internet Storm Center (ISC), malicious activities on both SSH (22) and MS-SQL (1433) ports increased significantly during 2006, which explains the trends seen here.

P2P and malicious connections reveal no diurnal patterns. Within Web traffic however, unsuccessful connection attempts account constantly for around 17.5% during all day, with exception of a drop to 10% during night time hours (2AM). We have no explanation for this phenomena other than HTTP connections are very rare in absolute number during night hours, which makes the statistical analysis more sensitive to behavior of individual applications or user groups.

Non-established connections: Non-established TCP connections have been further divided into connection attempts with one SYN packet only, attempts with direct RST reply and asymmetrical traffic (Fig.3). Due to transit traffic and hot-potato routing, 13% of the connections are asymmetrically routed. Naturally, it is not possible to observe a three-way handshake in this case.

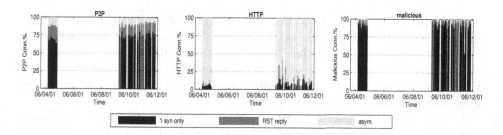

Fig. 3. Breakdown of non-established TCP connections

None of the traffic classes exhibits any significant diurnal pattern for non-established TCP connections. However, Fig.3 clearly highlights major differences between all three traffic classes. The already small fraction of non-established Web traffic (16.3% of all traffic) is mainly explained by asymmetrical traffic, and real unsuccessful connection attempts are very rare. Malicious traffic consists to a large degree of single SYN packet flows only. Single SYN flows are also dominating non-established P2P connections. While such connection attempts accounted for 71% in April, their fraction increase to 79% in November. This trend is also responsible for the increase of non-established P2P connections observed in Fig.2. Even if the high number of unsuccessful connection attempts within P2P traffic has been observed earlier [10], it is interesting to note that there is a clear trend in the fractions of one-SYN connections within P2P flows. The fraction increased by 23% (from 35% to 43%) within a period of 8 months.

Established Connections: Finally, established connections are broken down according to their termination behavior in Fig.4. Besides the proper closing approaches with one FIN in each direction or only one RST packet, as prescribed in the TCP standard, two unspecified termination behaviors have been observed. Connections closed by FIN, followed by an additional RST packet have been seen in direction of the initial SYN (typically the client) and the response (server). Finally, a number of connections were not closed during the measurement interval. The larger fraction of unclosed P2P connections is explained by the longer duration of P2P flows compared to Web traffic, as observed by Mori [7].

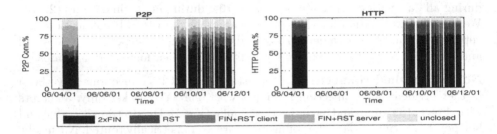

Fig. 4. Breakdown of established TCP connections

As for non-established connections, termination of Web connections neither shows significant trends nor diurnal patterns. HTTP connections are closed properly in 75% of all cases. Another 15% are closed by RST packets, mainly due to irregular web-server and browser implementations as noted by Arlitt [13]. FIN+RST behavior as well as unclosed connections (which corresponds to longer flows) are uncommon within Web traffic.

Even if there are no diurnal pattern observable, Fig. 4 indicates a significant change in termination behavior of P2P connections from spring to fall 2006. In April, only slightly less than half of the P2P connections have been closed properly with two FINs. As much as 20% of established P2P connections have been terminated with FIN plus an additional RST packet send by the server (or responding peer). A couple of popular hosts inside a student network have been identified as main source of this behavior. A commented text in the source code of a popular P2P client indicates that connections are closed with RST deliberately to avoid the TCP TIME_WAIT state in order to save CPU and memory overhead. In fall however, the fraction of FIN+RST terminations by the responder was reduced to around 8%, compensated by an increase in both valid TCP terminations, 2xFIN and single RST. Due to missing payload data, it was not possible to differentiate between different P2P software and version numbers. We suspect, that either the developers of the P2P application fixed this non-standard behavior in updated versions of the software, or the misbehaving P2P software lost popularity and was replaced by better behaving software by the users during 2006. However, the breakdown in Fig.4 shows that P2P traffic is mainly responsible for the large number of RST packets seen in todays networks.

5.2   Differences in Option Deployment

Finally, deployment of the most popular TCP options during connection established has been investigated for P2P and Web traffic (Table 1). For each of the
four most popular TCP options, three different possibilities are distinguished: established - the option usage was successfully negotiated in SYN and SYN/ACK
packets; neglected - the option usage was proposed in the SYN, but not included
in the SYN/ACK; and none - the option was not seen in the connection.

Table 1. Differences in TCP Option Deployment

(a) TCP Options in P2P Conn.

	MSS	SACK	WS	TS
estab.	99.9%	91.0%	14.9%	8.8%
neglected	0.1%	6.5%	0.6%	1.0%
none	0.0%	2.5%	84.5%	90.2%

(b) TCP Options in HTTP Conn.

	MSS	SACK	WS	TS
estab.	99.6%	65.7%	16.0%	13.4%
neglected	0.4%	27.9%	4.3%	4.3%
none	0.0%	6.4%	79.7%	82.3%

Option usage turned out to be remarkably constant, with neither longitudinal
nor diurnal trends. However, it is surprising to find such notable differences
in option usage between traffic classes, considering that protocol stacks in the
operating system, and not applications, decide about option usage. The MSS
option is almost fully deployed, which agrees with the fact that the MSS option
is set by default in all common operating systems. The SACK permitted option,
in fact also a default option, is commonly proposed by initiating hosts, but is in
28% of the Web connections neglected. Interestingly, this fraction is significantly
smaller in the case of P2P traffic, with only 6.5% neglecting SACK support.

While Linux hosts have the Window Scale (WS) and Timestamp (TS) options enabled by default, Windows XP does not actively use the options, but
replies with WS and TS when receiving SYN packets with the particular option.
This policy is well reflected by P2P connections, where WS and TS are rarely
neglected, but either established or not used at all. HTTP connections do not
really reflect this assumption, with 4.3% of WS and TS requests neglected by
servers. However, WS and TS are established more often within Web traffic.

We suspect that the usage of WS and TS options within P2P traffic some-
what reflects the proportions of Linux (WS and TS enabled by default) and
Windows systems (WS and TS disabled actively, but responding to request) on
the links measured. The differences in option deployment for Web traffic however stem from a differing communication nature. While Web traffic represents
classical client server communication, with one dedicated server involved, P2P
represents a loose network of regular user workstations. Web-servers, as a central
element, can thereby influence the behavior of larger numbers of connections.
This suspicion is further confirmed by the fact that a majority of the HTTP
connections neglecting usage of SACK are directed to less than 100 web-servers,
which consistently do not respond with SACK options. Such central elements
do not exist in P2P overlay networks. Furthermore, web-servers are more likely
to be customized or optimized due to their specific task, whereas user workstations usually keep default settings of the current operating system. Some active

measurement samples taken in October 2007 proved that popular web-servers, like google, yahoo and thePirateBay, still neglect SACK, WS or TS options.

6   Summary and Conclusions

In order to study trends and differences within the main traffic classes on the Internet, aggregated backbone traffic has been collected during two campaigns in spring and fall 2006 [11]. The collected packet level data has then been summarized on flow level. The resulting connections have finally been classified into P2P, Web and malicious traffic, using a connection pattern classification method [9]. An analysis revealed that overall traffic volumes are increasing for both TCP and UDP traffic, with highest activities at evenings. On diurnal basis, P2P and HTTP traffic exhibit different peak times. P2P traffic was found to be clearly dominating with 90% of the transfer volumes, especially during evening and night times. In contrast, HTTP traffic has its main activities (9% of the data-volumes) during office hours. Similar diurnal patterns have been observed in terms of connection numbers, even if P2P connections are not as dominating as in the case of data volumes. This indicates that P2P connections typically carry more data than Web traffic. Malicious and attack traffic is responsible for a substantial part of all TCP connections and UDP flows, but plays a minor role in terms of data volumes since it typically consists of 1-packet flows only. It was interesting to observe that the fraction of malicious TCP and UDP flows remained constant in absolute numbers both on diurnal and longitudinal basis, even though traffic volumes generally increased. This shows that malicious traffic (e.g. scanning attacks) forms a constant background noise on the Internet.

In terms of connection signaling behavior, major differences between the three traffic classes have been highlighted. The number of unsuccessful P2P connection attempts, which already dominated the P2P connection breakdown in spring, was shown to have increased further until fall. We conclude, that the large fraction (43%) of 1-packet flows on one hand and the large average data amounts per P2P connection on the other hand indicate a pronounced 'elephants and mice phenomenon' (Pareto principle) [7] within P2P flow sizes. Regarding termination behavior, P2P connections exhibit a clear trend towards higher fractions of proper closings in fall. HTTP connections on the other hand appear to behave comparable well according to specification at all times.

Finally, also TCP option deployment was shown to differ significantly between P2P and Web traffic. While P2P traffic rather reflects an expected behavior considering the default setting in popular operating systems, HTTP shows artifacts of the traditional client server pattern, with some dedicated web-servers neglecting negotiation for certain TCP options. This is especially true for the SACK option. We conclude that even though SACK is deployed by almost all P2P hosts and web-clients, a number of web-servers still neglect its usage. It is unclear to us, however, for which reasons web-server software or administrators would choose not to take advantage of certain TCP features, like SACK.

In the presented study, differences between traffic classes have been found in all aspects discussed, even if not always expected. The results provide researchers, developers and practitioners with novel, detailed knowledge about trends and influences of different traffic classes in current Internet traffic. The data analyzed was collected on a highly aggregated backbone link during a substantial time period, thus reflecting contemporary traffic behavior on one part of the Internet. Besides the general need of the networking and network security community to understand the nature of network traffic, information about behavior differences as seen 'in the wild' can be important when developing network applications, protocols or even network infrastructure. Furthermore, the results form valuable input for future simulation models.

References

[1] Moore, A.W., Papagiannaki, K.: Toward the Accurate Identification of Network Applications. In: Dovrolis, C. (ed.) PAM 2005. LNCS, vol. 3431, pp. 41–54. Springer, Heidelberg (2005)

[2] Sen, S., Spatscheck, O., Wang, D.: Accurate, scalable in-network identification of p2p traffic using application signatures. In: WWW 2004: Proceedings of the 13th Int. World Wide Web Conference, New York, USA (2004)

[3] Crotti, M., Dusi, M., Gringoli, F., Salgarelli, L.: Traffic classification through simple statistical fingerprinting. Computer Communication Review 37 (2007)

[4] Karagiannis, T., Broido, A., Faloutsos, M., Claffy, K.: Transport layer identification of p2p traffic. In: IMC 2004: Proceedings of the 4th ACM SIGCOMM conference on Internet measurement, Taormina, Sicily, Italy (2004)

[5] Gerber, A., Houle, J., Nguyen, H., Roughan, M., Sen, S.: P2p the gorilla in the cable. National Cable and Telecommunications Association (2003)

[6] Sen, S., Jia, W.: Analyzing peer-to-peer traffic across large networks. IEEE/ACM Transactions on Networking 12 (2004)

[7] Mori, T., Uchida, M., Goto, S.: Flow analysis of internet traffic: World wide web versus peer-to-peer. Systems and Computers in Japan 36 (2005)

[8] Perenyi, M., Trang Dinh, D., Gefferth, A., Molnar, S.: Identification and analysis of peer-to-peer traffic. Journal of Communications 1 (2006)

[9] John, W., Tafvelin, S.: Heuristics to classifiy internet backbone traffic based on connection patterns. In: ICOIN 2008: Proceedings of the 22nd International Conference on Information Networking, Busan, Korea (2008)

[10] Plissonneau, L., Costeux, J.L., Brown, P.: Analysis of peer-to-peer traffic on adsl. In: Dovrolis, C. (ed.) PAM 2005. LNCS, vol. 3431, pp. 69–82. Springer, Heidelberg (2005)

[11] John, W., Tafvelin, S.: (SUNET OC 192 Traces (collection)), http://imdc.datcat.org/collection/1-04L9-9=SUNET+OC+192+Traces

[12] John, W., Tafvelin, S.: Analysis of internet backbone traffic and header anomalies observed. In: IMC 2007: Proceedings of the 7th ACM SIGCOMM conference on Internet measurement, San Diego, CA, USA (2007)

[13] Arlitt, M., Williamson, C.: An analysis of tcp reset behaviour on the internet. Computer Communication Review 35 (2005)

The Cubicle vs. The Coffee Shop: Behavioral Modes in Enterprise End-Users

Frédéric Giroire[1], Jaideep Chandrashekar[2], Gianluca Iannaccone[2],
Konstantina Papagiannaki[2], Eve M. Schooler[2], and Nina Taft[2]

[1] INRIA, France
frederic.giroire@inria.fr
[2] Intel Research
first.initial.last@intel.com

Abstract. Traditionally, user traffic profiling is performed by analyzing traffic traces collected on behalf of the user at aggregation points located in the middle of the network. However, the modern enterprise network has a highly mobile population that frequently moves in and out of its physical perimeter. Thus an in-the-network monitor is unlikely to capture full user activity traces when users move outside the enterprise perimeter. The distinct environments, such as the cubicle and the coffee shop (among others), that users visit, may each pose different constraints and lead to varied behavioral modes. It is thus important to ask: is the profile of a user constructed in one environment representative of the same user in another environment?

In this paper, we answer in the negative for the mobile population of an enterprise. Using real corporate traces collected at nearly 400 end-hosts for approximately 5 weeks, we study how end-host usage differs across three environments: inside the enterprise, outside the enterprise but using a VPN, and entirely outside the enterprise network. Within these environments, we examine three types of features: (i) environment lifetimes, (ii) relative usage statistics of network services, and (iii) outlier detection thresholds as used for anomaly detection. We find significant diversity in end-host behavior across environments for many features, thus indicating that profiles computed for a user in one environment yield inaccurate representations of the same user in a different environment.

1 Introduction

Traditional studies of end-user behavior in a network typically have employed traffic traces collected from network aggregation points (routers, switches, firewalls, etc.). In modern enterprise networks, a large sub-population is mobile; laptop users move seamlessly in and out of the corporate office daily. When outside, the end-hosts are used in a number of places such as homes, airport lounges, coffee shops, etc. The VPN infrastructure of the enterprise ensures that users are never really cut-off from the resources on the corporate LAN. In fact, with the growing trend to support flexible telecommuting policies, and the ubiquity

M. Claypool and S. Uhlig (Eds.): PAM 2008, LNCS 4979, pp. 202–211, 2008.

of network connectivity while outside the corporate network, users spend fewer hours physically within the office cubicle, or at least within a single work locale.

Usage models are quite different inside and outside the office for a variety of reasons. Infrastructure services (email, directory, and print services) may simply be unavailable when users are outside the enterprise. Furthermore, locations outside the enterprise often have noticeable resource limitations (less bandwidth, less security, et cetera). Thus, users may be hampered from listening to streaming music, or may be wary of checking bank accounts when in a coffee shop. Conversely, the corporate acceptable usage policy may prohibit peer-to-peer file sharing applications on the corporate LAN, whereas it may be a staple application at home.

Previous work on building user-based profiles, such as in [1,2,3,4,5], does not consider the modality of the end-host when it is outside the enterprise. We argue that the growing trend to work outside the office and the distinct "usage-models" across the different environments, renders the single-view profile of the end-host (like the one generated from enterprise measurements alone) incomplete. In this paper, we explore the hypothesis that a single (static) profile for an end-host is inconsistent and/or incomplete. This has important consequences across the domains of enterprise security, network design, capacity planning and provisioning.

We analyze detailed traffic traces from a real corporate enterprise, where the traces were collected on the end-hosts themselves. This is in stark contrast to previous enterprise studies based on aggregate traffic, such as in [6,7]. With these traces, we quantify the differences in behavior of the individual end-hosts across three different environments in which they operate: (i) inside the corporate enterprise, (ii) outside but connected through the corporate VPN, and (iii) outside, meaning disconnected from the enterprise altogether. To the best of our knowledge, this dataset is the first to capture traffic at end-hosts themselves. By collecting traces in-situ, rather than in network, we are able to correctly track a host's traffic even when its address, location, and/or network interface changes - avoiding the difficulties posed by DHCP address changes and host mobility that can thwart the accuracy of in-network traffic traces.

In this initial exploration of the "environment diversity" hypothesis, we focus on three distinct types of features. These are (i) the median duration of a user's presence in each environment, (ii) the relative usage of network services (destination IP ports) per environment for end-hosts, and (iii) outlier detection thresholds (the 95th percentile) for TCP/UDP/ICMP connection counts as used by anomaly detection.

The contributions in this paper improve and clarify our understanding of end-host user profiles. Although our central hypothesis, i.e., that profiles need to change across environments, seems obvious, there has been no previous research quantifying such a hypothesis. This is most likely due to lack of availability of the right kind of data for such a study. This paper aims to explore this gap in end-host traffic characterization.

2   Data Description

Our dataset consists of packet traces collected at nearly 400 enterprise end-hosts (5% desktops and the rest, laptops) spanning approximately 5 weeks. A novel aspect to these traces is that they were collected *on* the individual end-hosts; this provides visibility into the end-host's traffic even as it leaves the office environment. Participants in our data trace collection were geographically distributed; 73% of the users were from the United States, 13% from Asia, 11% from Europe, less than 1% in each of Israel, Ireland and Latin America. All but a few users were based out of large offices in metropolitan areas. All the hosts in the study ran a corporate standard build of Windows XP. We solicited employees to sign up on a voluntary basis for the trace collection via organizational mailing lists, newsletters, and so forth. Cash prizes were offered as an added incentive to participate. Participants explicitly downloaded and installed the data collection software on their personal machines, thereby giving consent. We estimate that approximately 4000 employees were solicited, out of which approximately 1 in 10 installed the software. Overall, the data collection effort yielded approximately 400 GB of traces.

The collection software was written as a wrapper around the windump tool that logs packets in the well-known pcap format. The wrapper tracked changes in IP address, interface, or environment; upon such a change, windump was restarted and a new tracefile created. Importantly, every trace file was annotated with flags indicating the active network interface, the environment and if the logical VPN interface was active. Once installed, the software ran continuously (when the machine was on) for 5 weeks. For some users, it ran a few days less as they did not install the software immediately. Corporate policy strongly discourages the use of P2P applications, and hence our set of users is unlikely to be using any such software, even when outside the corporate environment.

To mitigate privacy concerns, we only collected the first 150 bytes of each packet. We did this simply to be able to infer the actual external destination when the packets went through the corporate proxy server. After identifying the actual destination, the payloads were discarded and only the packet headers retained. The post processing was carried out on a central server where traces were periodically uploaded. Moreover, all naming information regarding the user identity or machine identity was discarded upon upload of the traces. All solicitation emails contained a complete description of the data to be collected, the anonymizing procedures, and a disclosure of how the data was intended to be used. Because of this anonymization, we cannot know which traces came from engineers, managers, executives, etc.

Importantly, *all* the end-hosts in the study were *personally issued*, i.e., there is a single user per host. This is because in our corporation, each employee is given one laptop as their primary computer. Some employees, as needed, are additionally issued desktops; these are primarily used for running tests, simulations, etc. Most employees take their laptops home with them in the evening. Based on anecdotal evidence, employees generally shy away from allowing family members or others to use their computers. Hence we expect that the majority of our end-hosts have a single user, even when outside the corporate environment. Although a single user may use multiple machines, our intent here is *not* to characterize all aspects of the user at all times. Instead the focus is on all aspects of how a user uses a particular machine. This is what impacts whether or not a single machine should switch profiles as it, together with a user, moves between environments. In that sense, it does not matter what the user does with other machines.

3   Diversity Across Environments

Users move between three different environments– that we call `inside`, `vpn`, and `outside`. In the first, `inside` (the corporate network), the end-host is plugged into the office LAN almost always with a wired ethernet connection (on occasion connecting to the wireless LAN). In our enterprise, employees use laptops as their primary computer system, and while at work, these move between a docking station (at their desk), meeting rooms, corporate cafeterias, etc. In the `vpn` environment, users launch a VPN client that "logically" connects them into the office LAN. Note that here, users could be outside the office (the common case), or inside where they are on an unsecured wireless network, which exists solely as a gateway to the VPN, and cannot be used to reach the outside. Finally, when `outside`, the user is physically outside the enterprise network, and does not have any access to any of the enterprise infrastructure services (email, file & print server, etc.).

As an initial glance into our data, we show the movements of two users between these environments. In Fig. 1, we show a three week timeline. Here, the width of the contiguous blocks denote occupancy in that environment. First, we observe that both users actually use all three distinct environments. Although not shown here, due to lack of space, this is true for the vast majority of users (there were very few exceptions). Second, we note that these two users have very different behaviors in terms of how much time they spend in each environment, and how frequently they switch between environments. The user on the right is primarily in `vpn`, indicating that he may travel considerably or work from home. This user also tends to leave his VPN connection open during much of the weekend. This could indicate one of two things, either our user is someone who wants to be able to respond quickly when email arrives, or someone who perceives (as is common) that his computer is safer when the VPN is active. In contrast, the user on the left seems to have a more traditional work and leisure time pattern, using the `inside` mode during daytime on weekdays, the `vpn` mode in the evening on weekdays, and the `outside` mode on weekends. Clearly the `outside` mode for this user is likely to capture non-office related activities.

What is obvious here is that different users have different needs, at different times, to access the resources on the enterprise network. Aside from diversity across users, it is also natural to expect that a single user carries out different activities in the different modes. We now explore such behavior for a variety of measures.

206   F. Giroire et al.

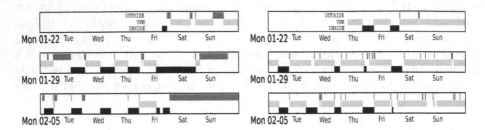

Fig. 1. A Tale of Two Users: time-line of two end-hosts over a 3 week window in the trace collection period

3.1   Environment Lifetimes

Motivated by Fig. 1, we first ask how much time a user spends in each environment. We define *environment lifetime* as the duration of contiguous time a user spends in a particular environment before changing it, restarting the machine or making it hibernate. Studying this statistic is key to solving many network design and planning problems. For instance, if one could model the time spent by users logged onto the VPN, the network operators could provision the VPN lines efficiently.

Fig. 2 is a set of scatter plots: each of these plots the median environment lifetime for individual users for two environments. In figure 2(a), each (x, y) point corresponds to a single user: the x value is the median time for `inside`, and y corresponds to `outside`. Similarly, Fig. 2(b) compares the lifetimes over `outside` and `vpn`, and finally, Fig. 2(c) compares `vpn` with `inside`. From these figures, it is quite clear that there is a marked difference in how long, in a single sittting, a user stays in each of these environments. Not surprisingly, for the most part, users spend more time `inside` as compared to the other two modes. It is interesting to see how short the environment lifetimes typically are for the `outside` mode. The lifetime spent `outside` can be anywhere from half to 10 times less than the typical lifetime for either the `inside` or `vpn` modes. An intuitive explanation for this could be that (i) the natural workday itself constitutes a window in which the employee is likely to stay in a single mode, and in addition (ii) when outside of work, the user's attention span (and time) is likely to be partitioned across mornings, evening, weekends, and interrupted by other domestic activities (meals, kids, etc.) which lead to shorter durations spent contiguously in the `outside` mode.

In comparing the environment lifetimes of `inside` mode versus `vpn` mode, we find interestingly that users exhibit tremendous diversity: some can stay on the `vpn` for 3 to 4 times as long as `inside`; others illustrate exactly the opposite behavior (points spread equally on both sides of the diagonal in Fig. 2(c). Users whose points lie near the extreme right side of this plot are likely to be employees who travel frequently, or who telecommute often, and thus their dominant work environment is through a VPN. We also observe, that even within the `inside` mode, there is great diversity across users - some have working sessions for 8 to 9

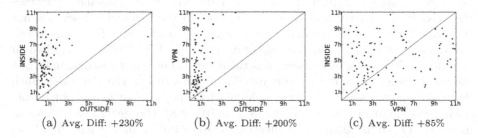

(a) Avg. Diff: +230%   (b) Avg. Diff: +200%   (c) Avg. Diff: +85%

Fig. 2. Median Lifetimes in different environments. Median values across users: outside=43min, vpn=3h45min, inside=6h30min.

hours, while for others the median time is 1 or 2 hours. The main takeaway from these statistics is that we see two kinds of diversity. There is tremendous diversity for each individual, in terms of the time the user stays pinned to particular environments. Not only do users spend vastly different amounts of time in each environment, but knowing a particular user's behavior does not reveal much about the others. Some users will have similar trends (regarding the fraction of time spent in each environment), whereas others exhibit completely opposite trends. We thus also see diversity across users for this measure.

3.2   Destination Port Diversity

We now examine whether there are quantitative differences in how network services are used in different environments. We use TCP and UDP destination ports as a useful proxy for "network service" (for the subset of ports we consider this is reasonable). Because it is impossible to exhaustively examine all destination ports, we focus on two logically formed groups. First, we study the ports associated with HTTP and web traffic (80,88,8080, 443) which we term the Web ports, and second, we look at the ports associated with Windows based services, that are popular in the enterprise (135,389,445,1025-1029), denoted MS Ports.

The particular metric of comparison that we use is the fraction of connections corresponding to a particular port (or group of ports). For every user and in each of the environments, we collect all the connections made to a particular port and the metric is computed as the *ratio of connections on that port to the total number of connections* (in the same environment). This is intended to capture a notion of what percent of a user's activities in each environment do they spend on a given service. Fig. 3 plots this metric for three different port sets, in each case comparing behavior across the inside and outside modes. In each scatter plot, a point corresponds to an individual user and the (x,y) coordinates are the connection fractions corresponding to inside and outside, respectively.

Fig. 3(a) plots the statistic for http traffic across the inside and outside environments (we exclude SSL traffic on port 443 from this plot and analyze that separately). The first thing we observe is the scattering of points over the entire graph. Importantly, nearly all these points are off the diagonal, indicating the percent of activity spent browsing the web in the two environments is not the same for users. Interestingly, there is no "typical user". For some users, the fraction of connections they generate that are HTTP is higher when in inside mode, and for other users, it is the reverse. The set of dense points along the x-axis indicates users that only use HTTP when inside the enterprise. Such users may have a second machine at home that they use for general browsing. Such users stand in contrast to the user at (0.1,0.8) who generates 8 times as many HTTP connections (as a function of his total traffic) when outside as opposed to when at work. This could capture a user that prefers to read news, or pursue other leisure activities, when outside the office.

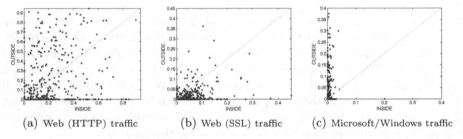

(a) Web (HTTP) traffic   (b) Web (SSL) traffic   (c) Microsoft/Windows traffic

Fig. 3. Comparing behavior across inside and outside environments

Similarly, we see in Fig. 3(b) for SSL traffic, that most of the points are off diagonal. Depending upon the user, the points can be a little or very far away from the diagonal. On most laptops, SSL constitutes a larger fraction of the total activity when the machine is inside the enterprise. In Fig. 3(c), we see a dramatic difference in the use of the MS ports. This is not surprising as many of these are primarily infrastructure services. The three plots confirm our hypotheses, that activity level profiles for a user are not the same in different environments. For some users, the differences may be small (but nonzero), whereas for others, the differences can be dramatic. We thus advocate that any profiling methodology that attempts to capture relative traffic measures — of a network service, for a given user on a particular machine — needs to be environment aware.

3.3   Thresholds on Behavioral Anomaly Detectors

Today, most enterprise end-hosts employ Host Intrusion Detection Systems (H-IDS) for security purposes. H-IDS systems typically include, among other things, a suite of anomaly detectors. From recent research, a popular approach to anomaly detection is to build behavioral profiles and use them to understand what is and isn't "normal" at an end-host. Many anomaly detectors define a threshold, [8,9,10], which defines the boundary between what is normal and abnormal for that host.

We now ask the important question as to whether or not such thresholds would vary for a given user, across different environments? If so, this would imply that

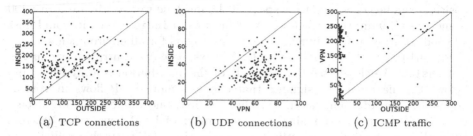

(a) TCP connections   (b) UDP connections   (c) ICMP traffic

Fig. 4. 95th %-ile values for *tcp, udp* and *icmp* protocols. Connection counts in 15 minute windows are used.

the configuration of anomaly detectors also needs to be environmentally-aware, possibility loading different profiles (i.e., thresholds) into the H-IDS, depending upon the current user environment.

Some detectors track the number of connections of a particular type within a time window. Here we will examine this type of feature for TCP connections, UDP connections and ICMP packet-pair counts. For these 3 protocols, we count the number of connections in 15 minute windows and build histograms for each user to indicate how many are likely. We compute thresholds that demarcate the 95th percentile point of these distributions, and consider these as the threshold value for the anomaly indicator. Considerable work has been devoted to the very specific problem of selecting suitable definitions of what constitutes an anomaly, or an outlier, however this topic is well outside the scope of this paper. Instead we pick a simple definition of an outlier and use it consistently across users and environments; this facilitates a straightforward comparison of the tail behavior of users across environments. Here, we study how this value differs across the environments.

In order to obtain connection records from raw packet traces, we use `bro` [11] to reassemble the flows/connections from the packet headers. The 95th percentile values for the three features are shown in Fig. 4. In each scatter plot, a point corresponds to the values, in the two environments, for an individual user.

The high-level observation is that points are considerably off diagonal in every case. Note that points on or near the diagonal correspond to users that have approximately the same threshold value in both environments being compared. For instance, take fig. 4(a): here most of the points are well off diagonal. Moreover, roughly half of the user population lies on either side of the diagonal. The latter hints at two user classes (of roughly equal population) for whom the value in one environment dominates. Take the point most extreme to the right: the 95%-ile corresponding to `outside` is 400, and 120 for `inside`. Thus, there is a higher intensity of outgoing TCP connections when in `outside`, while when in `inside`mode there are almost no 15 minute windows in which one sees more than 120 TCP connections. This is a marked difference. If a security anomaly detector tracking TCP connections, is configured with a threshold of 120/15min for all environments, then when the machine is `outside` a large number of false positives will be generated. Conversely, if the machine were configured with 400 connections/15min, then when the machine was in `inside`mode, it would miss all stealthy attacks. Clearly neither of these is good for all environments.

Fig. 4(b) plots the differences for UDP flows; here, we contrast usage in `vpn` and `inside`. We clearly see that the bulk of the distribution is away from (and below) the diagonal; this signifies that one sees more UDP flows in `vpn`, as opposed to `inside`. This seems puzzling at first; one would normally expect more traffic, and correspondingly larger number of UDP flows when `inside`. Upon closer inspection, we identified two destination UDP ports that contributed a large number of small sized flows; one was associated with the VPN client application and the other with a software compliance checker. The flows from these ports contributed significantly to the "rightward" skew of the points in Fig. 4(b). When the same plot was recomputed after filtering out flows from these two ports, the distribution of the points more closely resembled that in Fig. 4(a).

Finally, in Fig. 4(c) we compare ICMP traffic across `outside` and `vpn` environments. We see that there is very little ICMP traffic (to almost none) when the host is `outside`. Thus, ICMP traffic is extremely discriminating to the environment (more than TCP and UDP). This is possibly due to a lot of maintenance and network management traffic when the machine is on the VPN (a logical extension of the enterprise). This last observation strongly supports our hypothesis, i.e., that environment awareness is critical. A number of DDoS attacks, and some OS fingerprinting techniques make use of ICMP probes; large amounts of ICMP traffic are generally suspicious. In the figure, we see many users generate 200-300 ICMP packets within 15 minutes, and to be effective, an anomaly detection threshold would be set above this level. However, when we do this, we essentially provide a safe margin of the same amount (of ICMP traffic) when the host is `outside`; an infected or compromised machine could send out 200-300 ICMP packets without any fear of being flagged.

We conclude from this section, that because thresholds used by anomaly detectors define a boundary between normal and abnormal traffic, end-user based security mechanisms need to be designed to be "environment-aware". This is because these boundaries **do** change across environments for the same user.

4   Conclusion

Our study of common user-behavior features illustrates that most users exhibit significant diversity in how they use their machines in different environments. We show this on traces collected from end-hosts in an actual enterprise network. Regardless of whether we are looking at time spent in an environment, volumes of connections, http traffic, fraction of connections for Microsoft/Windows services, the measure can differ by anywhere from twice to 10 times as much in one environment as compared to another. These results illustrate that a profile computed in one environment will yield an inaccurate representation of user activity levels in another environment, for the majority of the users.

We showed how this could impact the configuration of anomaly detectors in H-IDS systems. These findings have implications for a number of other applications as well, such as resource allocation, VPN tunneling, and even virtual machine configurations. For example, if tomorrow's laptops employ different virtual machines for the home and work environments (such as the "red/green" VM proposal in [12]), then each VM should be configured to grab the appropriate profile before launching. We thus believe that "environmental awareness" is important for such applications. In the future we plan to study how some of these applications could be improved by using environmentally-aware profile information. We also plan to carry out user clustering to determine the minimal number of common profiles that could be used to capture the entire set of user behaviors.

References

1. McDaniel, P., Sen, S., Spatscheck, O., der Merwe, J.V., Aiello, B., Kalmanek, C.: Enterprise security: A community of interest based approach. In: Proc. of Network and Distributed System Security (NDSS) (Feburary 2006)
2. Tan, G., Poletto, M., Guttag, J., Kaashoek, F.: Role classification of hosts within enterprise networks based on connection patterns. In: Proc. of the USENIX Annual Technical Conference 2003, USENIX, pp. 2–2 (2003)
3. Karagiannis, T., Papagiannaki, K., Taft, N., Faloutsos, M.: Profiling the end host. In: Passive and Active Measurement, pp. 186–196 (2007)
4. Padmanabhan, V.N., Ramabhadran, S., Padhye, J.: Netprofiler: Profiling wide-area networks using peer cooperation. In: Castro, M., van Renesse, R. (eds.) IPTPS 2005. LNCS, vol. 3640, pp. 80–92. Springer, Heidelberg (2005)
5. Bhatti, N., Bouch, A., Kuchinsky, A.: Integrating user-perceived quality into web server design. In: Proc. of the 9th International World Wide Web conference on Computer networks, pp. 1–16. North-Holland Publishing Co, Amsterdam (2000)
6. Pang, R., Allman, M., Bennett, M., Lee, J., Paxson, V., Tierney, B.: A first look at modern enterprise traffic. In: Proc. of the Internet Measurement Conference (IMC), pp. 2–2. ACM, New York (2005)
7. Bahl, P., Chandra, R., Greenberg, A., Kandula, S., Maltz, D.A., Zhang, M.: Towards highly reliable enterprise network services via inference of multi-level dependencies. In: Proc. of ACM SIGCOMM, New York, USA, pp. 13–24. ACM, New York (2007)
8. Biles, S.: Detecting the unknown with snort and the statistical packet anomaly detection engine (SPADE) Computer Security Online Ltd
9. Jung, J., Paxson, V., Berger, A.W., Balakrishnan, H.: Fast portscan detection using sequential hypothesis testing. In: IEEE Symposium on Security and Privacy, p. 211 (2004)
10. Kreibich, C., Warfield, A., Crowcroft, J., Hand, S., Pratt, I.: Using Packet Symmetry to Curtail Malicious Traffic. In: Fourth Workshop on Hot Topics in Networks (HotNets-IV) (November 2005)
11. Paxson, V.: Bro: A system for detecting network intruders in real-time. Comput. Networks 31(23), 2435–2463 (1999)
12. England, P., Manferdelli, J.: Virtual machines for enterprise desktop security. Information Security Technical Report 11(4), 193–202 (2006)

A Two-Layered Anomaly Detection Technique Based on Multi-modal Flow Behavior Models

Marc Ph. Stoecklin[1], Jean-Yves Le Boudec[2], and Andreas Kind[1]

[1] IBM Zurich Research Laboratory
[2] Ecole Polytechnique Fédérale de Lausanne (EPFL)

Abstract. We present a novel technique to detect traffic anomalies based on network flow behavior in different traffic features. Based on the observation that a network has multiple behavior modes, we estimate the modes in each feature component and extract their model parameters during a learning phase. Observed network behavior is then compared to the baseline models by means of a two-layered distance computation: first, component-wise anomaly indices and second, a global anomaly index for each traffic feature enable effective detection of aberrant behavior. Our technique supports on-line detection and incorporation of administrator feedback and does not make use of explicit prior knowledge about normal and abnormal traffic. We expect benefits from the modeling and detection strategy chosen to reliably expose abnormal events of diverse nature at both detection layers while being resilient to seasonal effects. Experiments on simulated and real network traces confirm our expectations in detecting true anomalies without increasing the false positive rate. A comparison of our technique with entropy- and histogram-based approaches demonstrates its ability to reveal anomalies that disappear in the background noise of output signals from these techniques.

1 Introduction

Safeguarding availability and reliability of resources in computer networks poses a major challenge to network administrators. Conditions detrimental to a network's performance need to be detected in a timely and accurate manner. Such undesirable conditions are usually termed network anomalies; they include attacks and abuse of resources, significant changes of user behavior as well as failures of mission-critical servers and devices. Many of these conditions cannot be described by means of explicit signatures or differ slightly from known anomaly patterns though. Signature-based intrusion detection systems are thus likely to fail to detect them. Behavior-based anomaly detection techniques are a complementary approach to address these shortcomings. Their inherent assumption relies on the fact that anomalies are rarely observed in traffic and that if an abnormal condition is present, certain characteristics of the network behavior change. An anomaly-based detection system establishes baseline profiles of the normal behavior of a network and flags perturbations thereof as abnormal.

Anomaly detection systems operating at the network flow level have been widely discussed in the literature. In general, they assume that every traffic event leaves traces in distributions of flow level traffic features such as packet header fields (e.g., IP addresses, port numbers) and flow properties (e.g., the number of packets and octets transmitted,

M. Claypool and S. Uhlig (Eds.): PAM 2008, LNCS 4979, pp. 212–221, 2008.

flow duration). A feature distribution consists of the (normalized) number of flows observed in each component of the feature during a time interval. Many existing techniques apply a pre-processing step to a distribution (e.g., take its sample entropy) to obtain an estimate of its properties. However, precious information may be lost during the pre-processing before being presented to detection algorithms. Each component of a feature distribution is subject to variation and may exhibit multiple normal behavior modes (e.g., depending on time of day). Early summarization of feature distributions is therefore likely to miss such individual behavior patterns.

In this paper, we propose a flow-based technique to perform anomaly detection on two abstraction layers by taking the dynamic nature of individual components of traffic features into account. Our technique does not incorporate prior knowledge of normal and abnormal traffic characteristics and is therefore not bound to detect specific network anomalies. Instead, it makes use of positive learning samples to mine for normal behavior modes and to extract multi-modal model parameters in each component in an unsupervised manner. To compare observed network traffic with the learned modes, we propose a non-linear correlation system that finds the best matching model in each component. The resulting component-wise distances permit a twofold detection: (i) component-wise anomaly indices and, by aggregating the distances, (ii) a global index for each traffic feature. This duality enables detection of anomalies that affect isolated (e.g., host failures, DoS attacks) as well as multiple components (e.g., network scans, worm outbreaks). The modeling technique supports near real-time anomaly detection and on-line incorporation of administrator feedback to gradually reduce the false positive rate. To classify anomalies on demand, in-depth analysis of suspicious events is enabled by providing deviation vectors of the traffic features for decision support.

We validate our detection technique on real network traces collected in two production networks and a simulated dataset. We compare the technique experimentally with entropy- and histogram-based detection approaches and demonstrate its ability to detect anomalies that affect isolated components more reliably, thanks to late summarization in the process chain and individual component modeling. Thus, we believe that our technique can be a valuable component of a global detection system.

The remainder of this paper is organized as follows. Section 2 reviews related work and positions our contribution. In Sect. 3 we report our observations that led us to this approach. The anomaly detection technique is presented and discussed in detail in Sect. 4. In Sect. 5 we present the results of the evaluation and comparison.

2 Related Work

Anomaly detection has been extensively studied in recent years. Most prior work examines network traffic for specific anomaly patterns such DoS attacks, worm outbreaks, and network scanning and flooding. Other proposed techniques analyze overall traffic volume behavior, e.g., by applying edge detection [1], wavelet based signal analysis [2], or forecasting techniques [3]. Based on the assumption that anomalies are reflected as significant changes in traffic volumes, they flag peaks and shifts in volumes as suspicious events. Traffic volumes, however, comprise natural bursts and variability that are due to legitimate applications (e.g., backups, update rollouts, distributed computing), and therefore these approaches are likely to generate many false positives.

Lakhina et al. [4] showed that, due to the intrinsic low dimensionality of network flows, IP header feature measurements can be separated into disjoint subspaces that represent normal and abnormal behavior. They apply an information-theoretic analysis on feature distributions of IP addresses and service ports using entropy as an estimator of the characteristics [5]. Entropy describes the concentration and dispersal of a distribution in a single number and is a useful indicator for many traffic anomalies. However, entropy lacks the ability to discern differing distributions that possess the same amount of uncertainty. Also, divergence from a baseline value in single components may have little impact on the distribution's entropy and component-wise baselining is not applicable to entropy-based approaches. Consequently, observed network behavior may significantly deviate from usual behavior without being reflected by entropy.

Closer to our work, some approaches circumvent this shortcoming by comparing observed distributions to baseline distributions. Gu et al. [6] use relative entropy as a comparison metric with a single baseline and divide observed packets into classes according to layer 4 protocols, service ports, and a selection of particular TCP flags (SYN and RST). Our approach is agnostic with respect to characteristics that are prevalent during anomalies and does not use histogram-based traffic models. Hence, we expect to detect a broader set of network anomalies and higher sensitivity in identifying deviating behavior in individual components. Venkataraman et al. [7] developed a framework to transform arbitrary types of data sources (SNMP measurements, syslog output, NetFlow data) into constant-spaced real-valued time series. During the learning phase, they infer model parameters from the time series based on a range of assumptions (e.g., maximum and average value, percentile, etc.) and determine their confidence. In our work, we focus on time series of vectors with multiple components per feature, instead of one-dimensional time series, and assume multi-modal behavior patterns in each of them. We employ a two-layered detection technique that flags anomalies which affect both isolated components and multiple components in a feature.

3   Background and Motivation

Flow records exported by traffic meters (e.g., routers, switches) provide a large set of statistics of observed network flows. The statistics relate to different traffic *features* that include, for example, IP addresses, service ports, the number of packets and bytes, TCP flags, and start and end time of a flow. Each feature consists of a set of associated *components*, i.e., the actual values the statistics take. For example, port numbers 80/http and 22/ssh are components of the "service port" feature. When collecting flow records over a period of time, the total number of flows observed in each component can be counted and represented in a vector. We call such a "snapshot" of network traffic a *flow-count histogram*[1]. Figure 1 depicts a flow-count histogram representing the usage of service ports 1–200 observed in a production network during a 5-min period.

By monitoring a network over a long period of time, a time series of flow-count histograms can be collected for a given traffic feature; each histogram is a statistic

---

[1] We use the term histogram because continuous-type features are separated into bins.

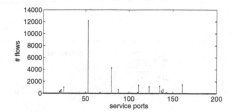

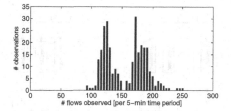

Fig. 1. Flow-count histogram of service ports 1–200 in one 5-min period

Fig. 2. Distribution of flow counts in 5-min periods for service port 22/ssh over one day

of the observed traffic over a sub-period of fixed duration. Instead of focusing on the histograms, we are now interested in the time series of flow counts in each component or, more precisely, the corresponding frequency distribution of the counts, see Fig. 2. By analyzing these flow-count distributions, we observed that the components generally exhibit multiple *behavior modes* depending, for example, on time of day, maintenance operations, or application and protocol states. Based on this observation, our anomaly detection technique extracts and models these behavior modes in each component.

We realized that anomalies affect the flow counts in one or more components of different traffic features and cause deviations from the models. For example, a worm outbreak may increase the flow counts in several service ports and IP addresses whereas a server outage leads to a decrease of the server's IP address flow counts.

4 The Detection and Learning Technique

Our technique consists of two parts: a learning phase and a detection phase. Both phases in turn are composed of two steps. In the *learning phase*, positive (anomaly-free) training data is analyzed and model parameters of the behavior are extracted in an unsupervised fashion. Then, based on the learned models, the detection logic is trained. In the *detection phase*, observed network behavior is compared to the baseline models and a detection operation is performed.

In general, acquiring entirely anomaly-free data in the learning phase is impractical; we therefore assume that the training data may contain a few anomalies. Consequently, we strive for a model extraction algorithm that is robust to the presence of a small fraction of deviating observations.

4.1 Notation

The detection technique operates on a set $\mathcal{F}$ of selected network features. Each feature $f \in \mathcal{F}$ consists of a finite number of components c_i^f with $i = 1, \ldots, n^f$. The value n^f represents the dimensionality of the feature space of f. For the sake of simplicity and without loss of generality, we will restrict ourselves to a single feature f henceforth and omit superscripts. A vector $\boldsymbol{h} = (h_1, \ldots, h_n)$ denotes a flow-count histogram where h_i is a non-negative count of flows associated to component c_i. Each component c_i has a baseline set $\Theta_i = \{\theta_1, \ldots, \theta_k\}$ that represents its learned behavior modes. A behavior mode θ_j is expressed by two model parameters: a mean value m_j of the observed flow counts in the mode and a scaling factor s_j reflecting their spread around m_j.

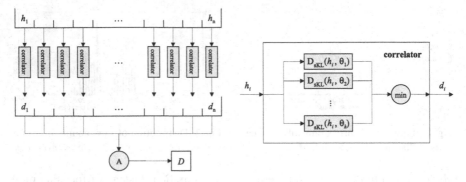

Fig. 3. Left: Computation of component-wise distances and a global distance of a flow-count histogram $h = (h_1, \ldots, h_n)$ at time t in a given traffic feature (left). Right: A correlator element selects d_i as the distance to the closest behavior mode $\theta \in \Theta_i$ with respect to h_i in component i.

4.2   The Detection Phase: A Non-linear Dual-Layered Detection Technique

Flow information is collected and divided into disjoint time intervals of fixed length ΔT. For each interval t, a flow-count histogram h is constructed. The processing of a histogram h begins with a matching operation performed in each component of h individually as depicted in Fig. 3 on the left. A non-linear correlator element (on the right) takes the flow count h_i in component i as an input and outputs the distance d_i to the closest known baseline model in the baseline set Θ_i. We employ a symmetrized version of the Kullback-Leibler (KL) distance that we found to have desirable properties in exposing anomalies; particularly decrease of flow counts with respect to the baseline value (e.g., in case of a failure) is penalized aggressively because of the asymmetry of the distance function around the baseline. The distance d_i between the observation h_i and the baseline set Θ_i is defined as $d_i = \min_{\theta \in \Theta_i} D_{\mathrm{sKL}}(h_i, \theta) = \min_{\{m,s\} \in \Theta_i} \frac{1}{s}(h_i - m) \log \frac{h_i}{m}$. In each component i of f, the anomaly index distance d_i is compared to a threshold value τ_i that has been derived during the learning phase. If d_i exceeds τ_i, then an abnormal deviation has been found in component i.

The component-wise distances d_i form an n-dimensional distance vector in the feature at time t; we term this data structure a *deviation vector*. It acts as an input to a summarization function A that calculates a global distance $D = A(d_1, \ldots, d_n)$ over all component-wise distances in feature f. We compute a weighted sum of the d_i, using the inverse of the thresholds τ_i as weights, to normalize the average contribution of each component. The index D is compared to the global threshold T for f. Aberrant behavior in the network feature f at time t is flagged if D exceeds T.

4.3   The Learning Phase: Training and Testing of Baseline Behavior Modes

The training data is split into two datasets: a training and a testing set. In a first learning step, our technique searches for behavior modes in the training set and extracts their model parameters. Then, the models are tested against the testing set to derive meaningful threshold values.

A natural way to derive models for flow-count distributions (see Fig. 2) is to consider them as the outcome of a finite set of random variables generating multi-modal data. In the literature, this is referred to as finite mixture models, e.g., Gaussian mixture models if the variables were normal, and the Expectation Maximization (EM) algorithm is commonly applied to estimate the parameters. However, the EM algorithm has two drawbacks with respect to our requirements: it is based on prior knowledge of the number of parameters to be found and, by attempting to find the best global fit of the data, it takes potential outliers into account.

We chose a deterministic technique that mines for local maxima in flow-count densities by means of a filtering approach, inspired by a density-based clustering algorithm that is robust to noise [8]. Our adapted algorithm iteratively scans the distribution, starting from low flow counts, and forms local groups of neighbors in regions of similar density. An imposed condition on the minimum number of observations per group, i.e., a small fraction g of the training data size, removes groups that are likely to be formed by abnormal events.

For each component i, we derive a threshold τ_i indicating the limit of acceptable deviation from the learned behavior modes. The first step of the detection phase (the distance computation) is run on the testing data to obtain the anomaly indices in each component. Then, while keeping in mind that some abnormal events may be present in the testing set, the $(1-g)$th percentile of the set of deviations is used as a heuristic for the thresholds. The global threshold T is computed in the same way as the component-wise thresholds, but on the global anomaly indices.

4.4 Discussion

In the first step of the detection phase, the technique computes the distances to the learned models and selects the smallest distance in each component individually. Implicitly, this refers to an on-the-fly construction of a dynamically composed baseline histogram with the closest known behavior mode placed in each of its component. Note that all learned models have the same weight in the selection process, independently of the number of observations in the learning phase.

We identified many benefits from operating on two abstraction layers. The *component-wise analysis* measures the coherence of the flow counts observed in each component i with the learned behavior modes in Θ_i. This indicator enables the detection of anomalies that affect single components, e.g., a host or service failure or the presence of an abnormally large number of flows with similar properties in certain features (e.g., Spam relaying, DoS attacks). However, some anomalies are likely to contribute only small deviations in many components and are therefore not detected at the component-wise detection layer. *Feature-wise analysis* is desirable to expose these deviations by accumulating the component-wise distances to a single global index. For example, a port scan will add a small increase in the flow counts of the ports scanned. While there is no substantial deviation from the normal behavior measured in each port, its presence is revealed by a large deviation in the global anomaly index of the ports.

Deviation vectors provide a detailed view on the measured deviations and enable interpretable analysis of suspicious activities to support operator decisions. By visually inspecting deviation vectors of features in which an alarm has been raised, the nature of the changes can be determined. Incorporation of administrator feedback, selective, and continuous model updates in the case of a false alarm is facilitated by the component-wise modeling strategy. The detection technique is suitable for on-line, near real-time anomaly detection, operating on traffic statistics from the preceeding time period.

5   Evaluation

We evaluate the technique on real and simulated network traces and compare it with entropy- and histogram-based approaches. We used three network traces in the evaluation: (i) Two weeks of NetFlow traces of internal and external traffic in an average-sized *production network*, (ii) ten days of NetFlow traces collected in a *data center*, and (iii) a publicly available simulated packet trace known as the DARPA Intrusion Detection evaluation 1999 dataset (*DARPA IDeval*) [9]. Even though we are aware of the many criticisms of the DARPA IDeval dataset in the literature, we use the labeled anomaly events to illustrate how the detection technique behaves in different network features. The NetFlow traces were stored in 5-minute intervals; we transformed the packet trace into the same format.

5.1   Results of the Evaluation

Production network. We used the first week of the traces as training data; two working days and one weekend day were selected as the training set, the remaining days as the testing set. As we do not assume the training data to be anomaly-free, we employ a robustness fraction of $g = 0.02$. By this, we enforce that each behavior mode consists of at least 18 (1.5 hours) from totally 864 observations in the training set. The features analyzed include service ports 1–1023, four IP class C subnets of critical server machines, the average packet size per flow, and TCP flags.

In Fig. 4, we depict the global deviations $D(t)$ measured in service ports and IP addresses, normalized by $\max_t D(t)$. A dashed line represents the global thresholds T in both features. In several time periods $D(t)$ exceeds T. On Thursday, for example, a large spike in both features is observed. The deviation vector in the service ports of that period, depicted in Fig. 5, shows a typical port scan pattern; in the IP address vector, a set of spikes indicates the machines scanned. On Tuesday, a deviation in several consecutive intervals is detected in an IP address component: a host's address appears significantly less frequently with respect to its baseline. The network administrator confirmed our observation of a two-hour outage of a mail server during that period.

Data center. The data center traces were exported by a router transferring more than 6 TB a day with average sending and receiving rates of 550 Mb/s and 100 Mb/s, respectively. Because of the homogeneous nature of the traffic mix in the data center, our main interest in these traces is the baselining of the behavior of individual server machines.

We trained the models for different IP address ranges from all days. A manual inspection of the flow-count distributions showed that the observed hosts generally exhibit 2 to 10 distinct behavior modes. The model extraction algorithm effectively determined these regions of high concentration; however, its conservative implementation extracted sometimes a few modes more than required.

DARPA IDeval. We used the "inside" network traces of the first week to establish the baseline models for service ports 1–1023, the 172.16.0.0/16 subnet, TCP flags, and

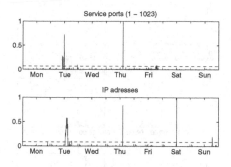

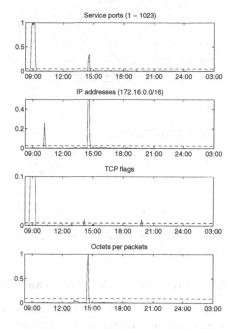

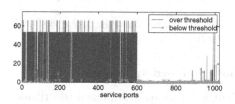

Fig. 4. Deviations in detection dataset of the "Production network" traces

Fig. 5. Deviation vector in service ports on Thursday at 14:20 ("Production network")

Fig. 6. Deviations on Tuesday, Mar 9, in the "DARPA IDeval" dataset

the average octet-per-packet ratio. As this data is known to be anomaly-free, we set $g = 0.001$, i.e., expecting 0.1 % of outliers. The advantage of the DARPA IDeval dataset is the availability of a "ground truth" of labeled anomalies for three weeks. We ran the detection phase on this data to verify whether the technique reliably reveals the network-related attacks and to evaluate how the anomalies affect the features considered.

In Fig. 6, the global anomaly indices for each feature on Tuesday (March 9) of the second week are depicted. While the port sweep attack on a single host starting at 08:44 is detected in the service ports and TCP flags (deviations in FIN and ACK/RST), it does not induce a significant global deviation in the IP addresses because of the moderate rate of about 1 scan per second. More interesting are the deviations at 14:25 in the service ports, IP addresses, and octets-per-packet ratios. Abnormally many flows use service ports 25/smtp and 53/dns and have specific octet-per-packet ratios. Our initial suspicion was confirmed by consulting the raw packets and the "ground truth": a vast flood of mail deliveries of similar size to a single recipient caused the mail server to look up the various sender host names at the DNS server. This event is labeled as "mail bomb" in the annotation.

All documented network-related attacks have been found with one repeating false positive event: a host tries to access a server on a closed port with several connection attempts in succession, generating deviations in the TCP flags SYN and ACK/RST. We suspect that this results from a possible application misconfiguration on the host.

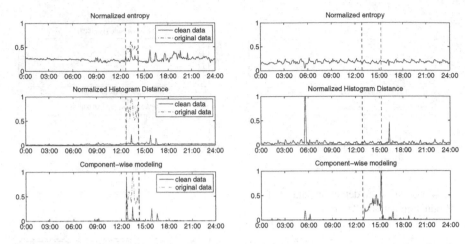

Fig. 7. Comparison of deviation signals for service ports of a port scan anomaly

Fig. 8. Comparison of the deviation signals for IP addresses of a server failure

5.2   Comparison with Entropy- and Histogram-Based Approaches

Entropy-based anomaly detection techniques quantify the nature of feature distributions (concentration or dispersal) by means of their information entropy. To mine for changes in time series of entropy values, various methods have recently been proposed, including PCA [5] and Kalman filters [10]. Histogram-based techniques extract representative feature distributions during a learning phase and the distance of an observed histogram to the closest learned model is analyzed in the detection phase. Ideally, each model describes a state of the network. In this section, we compare the output signals of the three techniques and use two traces from which we know that they contain abnormal conditions: (i) port scans on several hosts and (ii) a failure of a server.

Port scans. We constructed a "clean" trace by removing all activities generated by the scanner. By this, we obtain an ideal trace for learning purposes as well as for quantifying the deviation in the output signals. Figure 7 depicts the output signals of the service port distributions in the three approaches. In all approaches, a clear difference with respect to the "clean" trace can be observed during the scanning period. However, we observed that modeling of feature distributions in the histogram-based approach may entail increased variability in the background noise as a consequence of achieving a best fit over the entire feature. Component-wise modeling and dynamic histogram construction lead to more accurate baselines and reduce the false positive rate.

Server failure. The output signals of the three approaches for IP address distributions are depicted in Fig. 8. In the entropy- and histogram-based approaches, the change in the detection signal is barely noticeable. In the entropy case, we observed that changes of flow counts in a single component are weakly reflected in the distribution's properties and, thus, vanish in the normal variability of the entropy signal. The histogram-based approach's failure to distinguish the outage relates to a higher base distance that exists throughout the detection period in many components. Therefore, the absence of the server traffic adds only a small additional deviation to the output signal.

6 Conclusion

We have proposed an anomaly detection technique that takes the individual flow-count behavior of feature distribution components into account. The behavior modes are determined and corresponding models are extracted from assumed anomaly-free traces in an initial learning phase. In the detection phase, for every observed flow-count histogram of a feature, a two-layered distance computation is applied that constructs the best matching baseline histogram dynamically by placing the closest learned model parameters in each component. Deviation from these models is measured in each component as a component-wise index to detect anomalies that affect individual components. A global anomaly index of a feature is obtained by aggregating the distances over all its components; this index is a particularly useful indicator of aberrant traffic behavior affecting multiple components while generating only small deviations in each of them.

We showed that the detection technique flags various anomaly events with high precision and provides meaningful deviation vectors of the relevant time periods. A comparison with entropy-based approaches showed that with summarization of histograms during pre-processing valuable information about individual component behavior may be lost. In histogram-based modeling techniques, we saw that the state of a network cannot be captured with global feature distribution models. The proposed component-wise modeling strategy facilitates updates of the models and enables better interpretability of abnormal events to support decisions for network administrators.

References

1. Krishnamurthy, B., Sen, S., Zhang, Y., Chen, Y.: Sketch-based Change Detection: Methods, Evaluation, and Applications. In: ACM IMC 2003, pp. 234–247 (2003)
2. Barford, P., Kline, J., Plonka, D., Ron, A.: A Signal Analysis of Network Traffic Anomalies. In: Internet Measurement Workshop, pp. 71–82. ACM, New York (2002)
3. Brutlag, J.D.: Aberrant Behavior Detection in Time Series for Network Monitoring. In: LISA, pp. 139–146 (2000)
4. Lakhina, A., Crovella, M., Diot, C.: Diagnosing Network-wide Traffic Anomalies. In: ACM SIGCOMM 2004, pp. 219–230 (2004)
5. Lakhina, A., Crovella, M., Diot, C.: Mining Anomalies using Traffic Feature Distributions. In: ACM SIGCOMM 2005, pp. 217–228 (2005)
6. Gu, Y., McCallum, A., Towsley, D.F.: Detecting Anomalies in Network Traffic Using Maximum Entropy Estimation. In: ACM IMC 2005, pp. 345–350 (2005)
7. Venkataraman, S., Caballero, J., Song, D., Blum, A., Yates, J.: Black Box Anomaly Detection: Is It Utopian? In: Fifth Workshop on Hot Topics in Networks (HotNets-V) (2006)
8. Ester, M., Kriegel, H.P., Sander, J., Xu, X.: A Density-Based Algorithm for Discovering Clusters in Large Spatial Databases with Noise. In: ACM Conference on Knowledge Discovery and Data Mining (KDD), pp. 226–231 (1996)
9. Lippmann, R., Haines, J.W., Fried, D.J., Korba, J., Das, K.: The 1999 DARPA Off-line Intrusion Detection Evaluation. Computer Networks 34(4), 579–595 (2000)
10. Soule, A., Ringberg, H., Silveira, F., Rexford, J., Diot, C.: Detectability of Traffic Anomalies in Two Adjacent Networks. In: Uhlig, S., Papagiannaki, K., Bonaventure, O. (eds.) PAM 2007. LNCS, vol. 4427, pp. 22–31. Springer, Heidelberg (2007)

Malware in IEEE 802.11 Wireless Networks

Brett Stone-Gross[1], Christo Wilson[1], Kevin Almeroth[1], Elizabeth Belding[1],
Heather Zheng[1], and Konstantina Papagiannaki[2]

[1] Department of Computer Science,
University of California, Santa Barbara
{bstone,bowlin,almeroth,ebelding,htzheng}@cs.ucsb.edu
[2] Intel Research
Pittsburgh, PA
dina.papagiannaki@intel.com

Abstract. Malicious software (malware) is one of the largest threats
facing the Internet today. In recent years, malware has proliferated into
wireless LANs as these networks have grown in popularity and prevalence. Yet the actual effects of malware-related network traffic in open
wireless networks has never been examined. In this paper, we provide the
first study to quantify the characteristics of malware on wireless LANs.
We use data collected from the large wireless LAN deployment at the
67th IETF meeting in San Diego, California as a case study. The measurements in this paper demonstrate that even a single infected host can
have a dramatic impact on the performance of a wireless network.

1 Introduction

There has been ample research on the separate topics of malware and wireless
networks. A majority of malware research has focused on propagation modeling, detection, and application characterization [3][5][8]. The impact of malware
induced traffic on the performance of wired networks has been largely ignored,
because the effects of additional ingress and egress flows are mitigated by faster
access technologies and more bandwidth. However, limited resources in wireless
networks and the inherently broadcast nature of the medium creates valid concerns when considering network performance. This work analyzes these effects
which include MAC layer retransmissions, management frame collisions, and an
overall performance degradation due to increased congestion.

Wireless networks have been examined through experimental measurements
and simulations. Many studies have assessed wireless performance on deployed
networks [1][9][10][13]. Rodrig *et al.* captured wireless traffic and analyzed the efficiency of the 802.11 protocol [12]. They present how the efficiency significantly
degrades during periods of high contention with the majority of packets requiring link layer retransmissions due to packet loss and transmission errors. These
results are consistent with our own findings. Jardosh *et al.* examined methods
for detecting congestion in large-scale wireless networks [7]. They propose that
monitoring the channel busy time is a good measure of channel utilization. In

M. Claypool and S. Uhlig (Eds.): PAM 2008, LNCS 4979, pp. 222–231, 2008.

addition, network throughput and goodput can be used as metrics to identify congestion. Heusse *et al.* [6] found that anomalies in current multi-rate adaption algorithms of 802.11 cause an overall reduction in network performance, especially during periods of congestion. We also observed this behavior during several malware attacks. What all of these studies lack is an accounting of the extraneous packets that are injected into the network by malicious software.

We are the first to quantify, characterize, and correlate the effects of malicious network traffic on wireless performance. We believe that analyzing the effects malware can have on wireless networks is important. The applications of our research can lead to more realistic traffic models, justify the need for network protection, and improve the quality of service in wireless networks. In addition, recognizing these effects are beneficial in wireless network diagnostics [2][4].

The remainder of this paper is organized as follows. Section 2 describes our data collection and filtering process. In Section 3, the data sets are summarized. The effects that malware produced in the wireless network are examined in Section 4. Section 5 concludes with an overall summary of our findings.

2   Data Collection and Filtering

The wireless network deployed at the 67th IETF meeting was unusual due to both its large size and heavy utilization. The network provided an excellent opportunity to analyze the characteristics and prevalence of malware. With more than 1,700 unique users on the network, the resulting trace provided the equivalent of a small Internet Service Provider's (ISPs) perspective of malware attacks. Details of our data collection process at the IETF meeting and our subsequent malware identification process are discussed in this section.

2.1   Experimental Setup

The on-site network at the IETF meeting consisted of 30 802.11 a/b/g access points routed to a 44.7Mbps T3 backhaul link to the Internet. Participants utilized the Dynamic Host Configuration Protocol (DHCP) to obtain a publicly routable IP address in the 130.129/16 address range. No MAC layer encryption, Network Address Translation (NAT) devices, or firewalls were present in between the access points and the backhaul connection.

We collected data from two vantage points:

1. *Trunk Data Set*: Full data traces were recorded from a trunk mirror port on the router which managed the backhaul Internet link.
2. *Wireless Data Set*: Wireless sniffers were strategically positioned around the meeting near popular access points to record wireless traffic, as shown in Figure 1. Each wireless sniffer consisted of an IBM or Toshiba laptop with an Atheros chipset. Each sniffer was configured in RFMon mode to capture all management and data frames. Based on previous measurements [7], we estimate that each sniffer recorded more than 90% of frame transmissions.

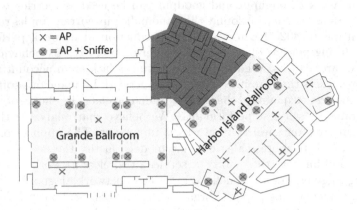

Fig. 1. Locations of wireless APs and data collection sniffers at the IETF meeting

Over 511 gigabytes of uncompressed data were collected at the trunk port along with another 131 gigabytes of uncompressed data recorded by the wireless sniffers. The data collected from the trunk port included some packets destined for a small on-site terminal room. This location was the only place in which attendees could access a wired Ethernet connection. We were able to identify traffic from the terminal room from the fixed set of IP addresses assigned by DHCP, by comparing IP addresses in both traces, and confirmed that less than 10% of the traffic observed in the trunk data set came from the terminal room.

2.2   Filtering Heuristics

In order to isolate malicious traffic from the normal flows present in the data set, we created a set of heuristic-based filters to detect abnormal behavior. We designed the filters around a set of assumptions about known malware behavior patterns, and then constructed an identification and measurement system. We observed that malware's traffic exhibits two primary types of traffic patterns:

- *Scanning behavior*: Worms and Trojans are typically spread by scanning large sequences of IP addresses on known ports. The scans search for vulnerable or weakly protected services (*e.g.*, default, weak or non-existent passwords) that can be exploited.
- *Flooding behavior*: Malware is often directed to attack other computers by flooding them with connection attempts (*e.g.*, a SYN flood).

One of the key characteristics of scanning behavior is that the machine in question will contact an abnormally large number of different IP addresses. This behavior will occur repeatedly on known vulnerable ports. Flooding behavior is best characterized as one machine initiating an unusually large number of connection attempts to one particular IP address.

For both behavior patterns, malicious traffic flows are often unidirectional and almost always short-lived. In the former pattern, scan attempts are often directed at unused IP addresses, or towards machines with firewalls which results in unidirectional traffic. SYN floods are by definition, unidirectional. If a scanner does manage to find a live target, it will attempt to either infect the host or guess the host's password, both of which are relatively brief affairs. Attempts may be repeated, but the connection is broken and reset each time, leading to bursty traffic flow characteristics. Another important consideration is that certain forms of malware including adware, keyloggers, and open relay proxies generate smaller amounts of network traffic and are consequently harder to identify. Therefore, the rest of our results should be considered as a lower bound of malware present.

3   Wireless and Trunk Data Analysis

Before we examined our wireless data set, we first developed a more general characterization of the network activity at the IETF using the trunk data set. Besides deriving network statistics, we used the trunk data set as the basis to identify malicious flows, which we later correlated with the more restricted data set obtained from the wireless sniffers.

3.1   Malicious Traffic Analysis

We begin by analyzing the malicious traffic present in the trunk data set. There were 109,740 unique external IP addresses in the trace, and 3,941 were implicated in malicious behavior, or about 3.6%. We identified 1,786 internal IP addresses, and out of this set 14 (0.8%) showed indications of malicious activity.

Overall, 272,480,816 egress TCP packets were sent over the course of the meeting, of which 4,076,412 (1.5%) were involved in malicious flows. 284,565,595 ingress TCP packets were received, of which 2,765,683 (1.0%) were malicious. In general these results appear consistent with a study by Kotz and Essien [9]. They recorded observing 0.9% of TCP traffic being sent to Microsoft RPC port 445, which they correlate with denial-of-service attacks against Windows 2000 machines. In our case, since we quantify scanning as well as flooding attacks across multiple services, our results represent a more complete view of overall malicious traffic percentages.

Although malicious TCP traffic accounted for an average of 1% of the total traffic at the IETF meeting, it accounts for a much larger percentage of TCP control traffic, defined as SYN and SYN-ACK packets. Thus, when data packets are not considered, the magnitude of malicious traffic becomes much more pronounced (as displayed in Figures 2 and 3). From this data, malicious flows are shown to account for a substantial portion of total TCP connection requests, occasionally rising above 50%. During a massive SSH password cracking attempt on Friday morning, nearly 100% of all TCP control traffic was part of the attack, and is clearly evident in Figures 2 and 3. In addition to conducting an analysis of malware behavior within the IETF network, we also attempted to isolate what

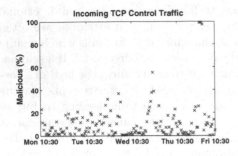
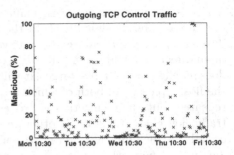

Fig. 2. Instantaneous percentage of incoming malicious TCP traffic

Fig. 3. Instantaneous percentage of outgoing malicious TCP traffic

effects such traffic had on the wireless medium itself. Although we were able to identify many attacks in the trunk data set, pinpointing these same attacks in the wireless data set proved to be difficult since our sniffers did not observe all wireless LAN traffic across all access points. From the set of malicious flows that were detectable in the wireless data sets, many proved unsuitable for analysis. The reasons include the following:

1. Ingress attacks that involved only a few total packets.
2. Egress scanning attacks which, though long lived, only generated a few packets per second.
3. Ingress port scans that were distributed over hosts on all 30 access points.
4. Backscatter from DoS attacks throughout the Internet that produced unsolicited TCP SYN ACKs, resets, and ICMP replies [11].

Although the preceding cases were not ideal for analyzing MAC characteristics, these attacks still had an overall effect as more than 1% of all packets were malicious and present in the wired and wireless data sets. The most substantial effects on wireless performance were produced by malicious flows that originated within the network. Therefore, we examined several of these egress flows under light and heavy channel utilization.

4   Quantifying the Impact of Malware

As previously discussed in Section 3.1, malicious egress flows were well suited for our analysis since these flows consumed more bandwidth, and caused more collisions than malicious ingress flows. In order to understand the impact of these malicious flows on the MAC layer, we aggregated statistics for channel utilization, throughput, probe requests/responses, data packets/retries/acknowledgments, and transmission rates. At the transport layer we computed the TCP Round-Trip-Times (RTT) to determine the end-to-end delay.

Table 1. The effects on TCP RTT of an ICMP flood and NetBIOS attack

	Non-Attack Interval	During Attack	Percent Increase
Avg (Egress)	64.7 ms	99.2 ms	53.23%
Avg (Ingress)	23.4 ms	36.1 ms	54.36%
Median (Egress)	41.6 ms	85.0 ms	104.33%
Median (Ingress)	3.2 ms	6.8 ms	112.50%

4.1 Malware Attacks in Wireless Networks

We performed a detailed analysis of two of the largest attacks occurring in the wireless data sets during the meeting based on packets per second and bandwidth. These types of attacks were also the most common that we observed. They included an ICMP ping flood combined with a NetBIOS exploit and a TCP SYN Flood.

ICMP Flood and NetBIOS Exploit. One of the largest network attacks observed during the entire meeting was an ICMP ping sweep across a range of IP addresses. The attack was used to probe for machines and prepare for a subsequent NetBIOS worm exploit. The malicious flow persisted for approximately 18 minutes and 7 seconds occurring late Thursday afternoon during the plenary session between 17:02:38 and 17:20:45. The attack created 79,289 packets at an average rate of 117 packets per second with a maximum burst of 235 packets per second. The impact of the flow drove the channel utilization to nearly 100%, and caused both a rise in the number of link layer data retries (retransmissions) and a reduction in the transmission rates (shown in Figure 4). The metric in Figure 4(b) shows the two primary ranges of transmission rates of 11-18Mbps and 48-54Mbps that were used by wireless clients. The rectangular regions in Figure 4 and 5 indicate the periods of malicious traffic flow.

As part of our analysis, we also discovered a brief period in the middle of the ping flood just after 17:09:00 when the attack halted. This temporary pause resulted in a reduction in utilization, an increase in data transmission rates, and fewer data retries. Unfortunately we were not able to determine why the attack was suspended during this two minute interval, but we conjecture that the infected machine may have become unresponsive and was rebooted.

An additional result that we observed in our analysis was that overall, the combined throughput on the channel remained relatively constant at 4,412 KB/s over the course of the attack. However, the average and median RTT increased by more than 50% and 100% respectively for all TCP flows. Table 1 displays the average and median RTTs for a 10 minute interval before and after the attack with respect to the RTT during the attack.

There are several conclusions that can be drawn based on these results. First, the attacker was not only able to adversely affect other clients' performance, but also obstruct the access point's probe responses to clients who were searching for access points via probe requests. This is evident in Figure 4(d), which illustrates the spike in probe responses immediately after the attack occurred.

Consequently, the attack exacerbated a problem in the wireless network in that probe requests and responses were essentially jammed during heavy utilization. Access point control packets such as beacons, probes, and other management frames were also lost or delayed, and therefore served no productive purpose and only contributed to the overall network congestion.

A reduction in client transmission rates occurred due to the Auto Rate Fallback (ARF) mechanism, as illustrated in Figure 4(b), due to increased packet loss. As a result, packet transmission times increased, which further increased the channel busy time. The purpose of ARF is to combat lossy channel conditions by sending data at lower rates (*i.e.*, provide more robust modulation and coding schemes), and thus decrease the likelihood that data is lost because of radio noise. However, using the ARF strategy is a poor choice in this case since dropped packets are due to packet collisions and not noise interference. During these congested periods, this behavior created a negative feedback loop as client queues filled, but were unable to effectively drain due to contention compounded by slower transfer rates. Therefore, the delay for each host increased as they continuously waited for the channel to become idle.

The dramatic increase in TCP delay, as shown in Table 1, can be attributed to the additional strain that this attack placed on the link layer. Accordingly, the attack produced a large amount of data retransmissions. During the attack nearly 25% of all MAC layer frames were retransmissions, and at the peak of the attack

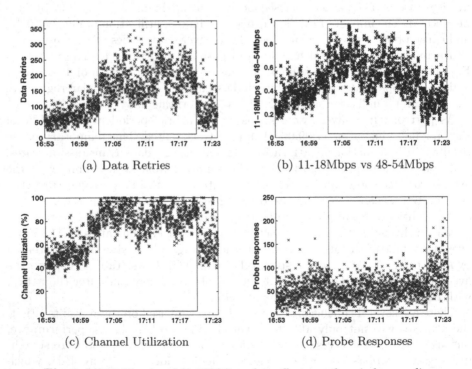

(a) Data Retries

(b) 11-18Mbps vs 48-54Mbps

(c) Channel Utilization

(d) Probe Responses

Fig. 4. ICMP Flood and NetBIOS exploit effects on the wireless medium

almost 50% of all packets were retransmissions. In addition, as clients following the ARF procedure reduced their transmission rates, the channel became even more congested as transmissions took longer to complete. These characteristics had a significant impact on TCP delay due to the fact that these MAC layer delays and losses were assumed to be caused by end-to-end congestion. Hence TCP transmission timeouts occurred, which reduced the congestion window.

TCP SYN Flood. Another one of the more obvious attacks that we observed was a TCP SYN flood directed at an external server on Port 80 involving over 6,000 connection requests. The attacker in question emitted three bursts of attack traffic that began Thursday afternoon at 12:59:57 and numbered up to 109 packets per second for 30 seconds.

Figure 5 combines several of these measurement metrics during the initial attack, which lasted for only 30 seconds. The peaks in the numbers of data packets correspond to periods of attack. As shown in Figure 5(c), the aggregate channel utilization for this particular access point, while elevated, was not near bottleneck limits. What was most impacted by the SYN flood was the data retry rate, which peaked in the midst of the attack. This result indicated a higher rate of contention and collisions at the MAC layer due to the attacker's rapid transmission of single SYN packets. The result was an increase in the overall end-to-end latency as the MAC layer struggled to reliably deliver packets. During this attack, the average RTT increased by more than 33% with 16% of all frames consisting of MAC layer retransmissions. At the peak of the attack, more than 30% of all frames were data retransmissions.

Additionally, the aggregate number of probe requests and probe responses to and from all access points increased during the initial attack as illustrated in Figures 5(b) and 5(d). This result indicates that the attacker may have aggravated existing hidden terminal problems, thereby causing collisions and data retries. This behavior then triggered nearby clients that were connected to the same access point to begin probing for other access points offering better connectivity. While these effects do not appear catastrophic, it is evident that the probe responses and data retries increased by more than twice their averages over regular traffic intervals. Analogous to the ICMP ping flood, the number of probe responses more than doubled immediately after the attack. This behavior occurred in response to the outstanding probe requests that were partially blocked during the attack interval.

4.2 Effects of Malicious Flows on Wireless Performance

Our findings show that the presence of active malware in a congested wireless network harms performance by reducing client transmission rates and increasing data retries. The results also demonstrate that the end-to-end delay for TCP connections rise commensurately with slower data rates and greater numbers of packet collisions. These effects would likely have a significant impact on real-time applications. Under heavy utilization, access point management frames can be obstructed and increase the delay in client handoffs, authentications, and

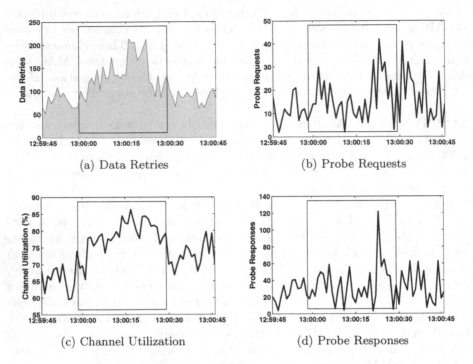

(a) Data Retries   (b) Probe Requests

(c) Channel Utilization   (d) Probe Responses

Fig. 5. TCP SYN flood effects on the wireless medium

associations, further degrading performance. By comparing the effects of the NetBIOS attack with the TCP SYN flood, we can determine that faster sending rates and larger packets have a more significant effect on the wireless medium since the channel is busy for longer periods of time. In addition, the 802.11 CSMA protocol worked well in preventing small TCP SYN packets from dominating the channel during malicious traffic flows.

5   Conclusion

The study of malware on wireless systems is becoming increasingly important as more devices communicate openly over-the-air. In this paper, we analyzed the effects that malware-driven attacks can have on 802.11 performance. The most severe consequence is an increase in RTTs, which can hinder real-time communication. Wireless quality of service is also virtually impossible without developing mechanisms to reduce unwanted link layer contention.

The results that we present are from single attackers' outgoing malware attacks. Left unabated, the prevalence of malware will lead to a higher concentration of attackers and potentially deny service to legitimate users. This makes the protection of connected machines an especially pertinent objective for wireless network operators. In addition, as worms and botnets become more sophisticated, we believe that the exploitation of wireless networks by mining sensitive information from unencrypted transmissions will become routine. Malware will also adapt to preserve its own anonymity by spoofing the source of attacks. Consequently, the effects of multiple compromised machines on a single wireless access point will become more significant as malware evolves to specifically exploit the wireless medium. Therefore, a lightweight solution will be essential to ensure optimal network performance and protect users' sensitive data.

References

1. Balachandran, A., Voelker, G.M., Bahl, P., Rangan, P.V.: Characterizing User Behavior and Network Performance in a Public Wireless LAN. In: Proc. of ACM SIGMETRICS, Marina Del Rey, CA, June 2002, pp. 195–205 (2002)
2. Chandra, R., Padmanabhan, V., Zhang, M.: WiFiProfiler: Cooperative Diagnosis in Wireless LANs. In: Proc. of MobiSys, Uppsala, Sweden (June 2006)
3. Chen, Z., Gao, L., Kwiat, K.: Modeling the Spread of Active Worms. In: Proc. of IEEE INFOCOM, San Francisco, CA (April 2003)
4. Cheng, Y., Afanasyev, M., Verkaik, P., Benko, P., Chiang, J., Snoeren, A., Savage, S., Voelker, G., Kwiat, K.: Automating Cross-Layer Diagnosis of Enterprise Wireless Networks. In: Proc. of ACM SIGCOMM, Kyoto, Japan (August 2007)
5. Gu, G., Porras, P., Yegneswaran, V., Fong, M., Lee, W.: BotHunter: Detecting Malware Infection Through IDS-Driven Dialog Correlation. In: Proc. of Usenix Security Symposium, Boston, MA (August 2007)
6. Heusse, M., Rousseau, F., Berger-Sabbatel, G., Duda, A.: Performance Anomaly of 802.11b. In: Proc. of IEEE INFOCOM, San Francisco, CA (March 2003)
7. Jardosh, A., Ramachandran, K., Almeroth, K., Belding-Royer, E.: Understanding Congestion in IEEE 802.11b wireless networks. In: Proc. of Internet Measurement Conference, Berkeley, CA (October 2005)
8. Kirda, E., Kruegel, C., Banks, G., Vigna, G., Kemmerer, R.: Behavior-based Spyware Detection. In: Proc. of Usenix Security Symposium, Vancouver, Canada (August 2006)
9. Kotz, D., Essien, K.: Analysis of a Campus-wide Wireless Network. In: Proc. of ACM MOBICOM, Atlanta, GA (September 2002)
10. Meng, X., Wong, S., Yuan, Y., Lu, S.: Characterizing Flows in Large Wireless Data Networks. In: Proc. of ACM MOBICOM, Philadelphia, PA (September 2004)
11. Moore, D., Voelker, G.M., Savage, S.: Inferring Internet Denial-of-Service Activity. In: Proc. of Usenix Security Symposium, Washington D.C (August 2001)
12. Rodrig, M., Reis, C., Mahajan, R., Wetherall, D., Zahorjan, J.: Measurement-based Characterization of 802.11 in a Hotspot Setting. In: Proc. of ACM SIGCOMM Workshop on Experimental Approaches to Wireless Network Design and Analysis (E-WIND), Philadelphia, PA (August 2005)
13. Schwab, D., Bunt, R.: Characterizing the Use of a Campus Wireless Network. In: Proc. of IEEE INFOCOM, Hong Kong, China (March 2004)

Author Index

Lecture Notes in Computer Science

Sublibrary 5: Computer Communication Networks and Telecommunications

Vol. 4269: R. State, S. van der Meer, D. O'Sullivan, T. Pfeifer (Eds.), Large Scale Management of Distributed Systems. XIII, 282 pages. 2006.

Vol. 4268: G. Parr, D. Malone, M. Ó Foghlú (Eds.), Autonomic Principles of IP Operations and Management. XIII, 237 pages. 2006.

Vol. 4267: A. Helmy, B. Jennings, L. Murphy, T. Pfeifer (Eds.), Autonomic Management of Mobile Multimedia Services. XIII, 257 pages. 2006.

Vol. 4240: S.E. Nikoletseas, J.D.P. Rolim (Eds.), Algorithmic Aspects of Wireless Sensor Networks. X, 217 pages. 2006.

Vol. 4238: Y.-T. Kim, M. Takano (Eds.), Management of Convergence Networks and Services. XVIII, 605 pages. 2006.

Vol. 4235: T. Erlebach (Ed.), Combinatorial and Algorithmic Aspects of Networking. VIII, 135 pages. 2006.

Vol. 4217: P. Cuenca, L. Orozco-Barbosa (Eds.), Personal Wireless Communications. XV, 532 pages. 2006.

Vol. 4195: D. Gaiti, G. Pujolle, E.S. Al-Shaer, K.L. Calvert, S. Dobson, G. Leduc, O. Martikainen (Eds.), Autonomic Networking. IX, 316 pages. 2006.

Vol. 4124: H. de Meer, J.P.G. Sterbenz (Eds.), Self-Organizing Systems. XIV, 261 pages. 2006.

Vol. 4104: T. Kunz, S.S. Ravi (Eds.), Ad-Hoc, Mobile, and Wireless Networks. XII, 474 pages. 2006.

Vol. 4074: M. Burmester, A. Yasinsac (Eds.), Secure Mobile Ad-hoc Networks and Sensors. X, 193 pages. 2006.

Vol. 4033: B. Stiller, P. Reichl, B. Tuffin (Eds.), Performability Has its Price. X, 103 pages. 2006.

Vol. 4026: P.B. Gibbons, T. Abdelzaher, J. Aspnes, R. Rao (Eds.), Distributed Computing in Sensor Systems. XIV, 566 pages. 2006.

Vol. 4003: Y. Koucheryavy, J. Harju, V.B. Iversen (Eds.), Next Generation Teletraffic and Wired/Wireless Advanced Networking. XVI, 582 pages. 2006.

Vol. 3996: A. Keller, J.-P. Martin-Flatin (Eds.), Self-Managed Networks, Systems, and Services. X, 185 pages. 2006.

Vol. 3976: F. Boavida, T. Plagemann, B. Stiller, C. Westphal, E. Monteiro (Eds.), NETWORKING 2006. Networking Technologies, Services, and Protocols; Performance of Computer and Communication Networks; Mobile and Wireless Communications Systems. XXVI, 1276 pages. 2006.

Vol. 3970: T. Braun, G. Carle, S. Fahmy, Y. Koucheryavy (Eds.), Wired/Wireless Internet Communications. XIV, 350 pages. 2006.

Vol. 3964: M.Ü. Uyar, A.Y. Duale, M.A. Fecko (Eds.), Testing of Communicating Systems. XI, 373 pages. 2006.

Vol. 3961: I. Chong, K. Kawahara (Eds.), Information Networking. XV, 998 pages. 2006.

Vol. 3912: G.J. Minden, K.L. Calvert, M. Solarski, M. Yamamoto (Eds.), Active Networks. VIII, 217 pages. 2007.

Vol. 3883: M. Cesana, L. Fratta (Eds.), Wireless Systems and Network Architectures in Next Generation Internet. IX, 281 pages. 2006.

Vol. 3868: K. Römer, H. Karl, F. Mattern (Eds.), Wireless Sensor Networks. XI, 342 pages. 2006.

Vol. 3854: I. Stavrakakis, M. Smirnov (Eds.), Autonomic Communication. XIII, 303 pages. 2006.

Vol. 3813: R. Molva, G. Tsudik, D. Westhoff (Eds.), Security and Privacy in Ad-hoc and Sensor Networks. VIII, 219 pages. 2005.

Vol. 3462: R. Boutaba, K.C. Almeroth, R. Puigjaner, S. Shen, J.P. Black (Eds.), NETWORKING 2005. XXX, 1483 pages. 2005.